WE DO THINGS DIFFERENTLY

WE DO THINGS DIFFERENTLY

DIFFERENTLY

THE OUTSIDERS REBOOTING

OUR WORLD

MARK STEVENSON

The Overlook Press
New York, NY

This edition first published in hardcover in the United States in 2018 by
The Overlook Press, Peter Mayer Publishers, Inc.

141 Wooster Street
New York, NY 10012
www.overlookpress.com

For bulk and special sales, please contact sales@overlookny.com,
or write us at the address above.

First published in Great Britain in 2017 by Profile Books

Cataloging-in-Publication Data is available from the Library of Congress

Manufactured in the United States of America
ISBN: 978-1-4683-1583-7

1 3 5 7 9 10 8 6 4 2

For Caroline and Emmett,
who help me think differently

'The future belongs to those who believe
in the beauty of their dreams.'
Eleanor Roosevelt

'There's a way to do it better. Find it.'
Thomas Edison

CONTENTS

INTRODUCTION
WINDMILLS

'Never interrupt someone doing what you said couldn't be done.' – AMELIA EARHART, AVIATOR

There is a man in the sleepy market town of Bishop's Stortford, Hertfordshire who has found a solution to two of humanity's biggest challenges – using only his lawnmower and a can of antifreeze. In Boston an engineer with no medical training has given the healthcare profession access to something more powerful than any drug ever created. Just outside the city of Ranchi, north-east India, a young man is growing crops in places that accepted wisdom would suggest that it's hopeless to farm, while in Brazil an idea first proposed by some neighbourhood activists is achieving something many would consider impossible – it makes politicians popular.

There have always been a subset of people who think differently. A smaller number *do* differently, people who look at the status quo and not only think 'I could fix that' but actually roll their sleeves up and start working.

And never have we needed them more.

We live in the eye of a storm, a time in history where humankind must change the way it organises itself or face disastrous consequences. Our energy and food systems are increasingly unsustainable, promising an entwined environmental, economic and humanitarian crisis of unprecedented proportions. Democracy, where it exists, is descending into alienating tribalism. Inequality is rife. If you're lucky enough to enjoy a free press, it's likely you don't trust it. The world's healthcare systems are, in reality, astonishingly expensive and labyrinthine *sick*-care systems. And in most parts of the developed world, our education systems still seem trapped in the last century.

It's easy to feel despondent. But for some individuals, the roll call of bad news (not helped by the fact that as far as the media is concerned the bad news is pretty much the *only* news) isn't a cause for despair, but a call to arms. I know because I spend my life hunting them out and trying to learn their lessons. When it comes to the future they are here to remind us that there are many more options available than the leaders of any corporation, political party, pressure group, religion, academic institution or media outlet would have you believe.

Such pioneers have never had it easy. In 1532 the maestro of change and original political scientist Niccolò Machiavelli's famous political treatise *The Prince* was published. In it he wrote:

> 'It ought to be remembered that there is nothing more difficult to take in hand, more perilous to conduct, or more uncertain in its success, than to take the lead in the introduction of a new order of things. Because the innovator has for enemies all those who have done well under the old conditions, and lukewarm defenders in those who may do well under the new. This coolness arises partly from fear of the opponents, who have the laws on their side, and partly from the incredulity of men, who do not readily believe in new things until they have had a long experience of them.'

In short, change might sound possible in principle but we'll only believe it if we can see it.

Welcome, then, to *We Do Things Differently* – a follow-up, or, more exactly a prequel, to my previous book, *An Optimist's Tour of the Future*. For this is not so much a book about the future as about the here and now. It charts a journey to find the people who, despite the resistance of those who benefit from the status quo, are putting brave and alternative futures on the table – new ways of organising ourselves that address the grand challenges of our age. It features innovators reshaping the education system, exploring new forms of government, reforming the world of healthcare and medicine, re-booting cities and changing the way we think about and produce our food and energy.

These innovators are not tinkering with the existing system, but looking to change the system itself.

Of course the world is replete with armchair sages telling us what they think the future should be like – and how much better it would be if we agreed with them. So, I had one over-riding criterion for inclusion in my itinerary. The innovators had to be *succeeding right now in the real world*. Whatever their idea, I wanted to be able to touch it, meet the people making and benefiting from it, see 'the steel in the ground' as the saying goes. It had to be working and I had to be able to see it working.

I travelled from the urban devastation of Detroit to a small town on the Austrian-Hungarian border; from the leading genetics labs in the world to one of the toughest housing estates in Britain. I met brilliant people from all walks of life, from poor farmers to hipster software geeks, from some of the world's highest-ranking scientists to a lone unqualified genius in a shed, from a nightclub owner turned headteacher to an international basketball ace turned engineer. It's a cast of characters who are by turns inspiring, demanding and driven – the pioneers, architects and builders of a surprising and hopeful future – albeit one that is presently below the radar. People who really do do things differently and invite us to do the same.

There is an old Chinese proverb:

'When the winds of change blow, some people build walls, others build windmills'.

This is a book about the windmills.

1 MY BROTHER'S KEEPER

'Beware the fury of a patient man.'
– JOHN DRYDEN, POET

It is with some trepidation that I approach a well-appointed Victorian house in the affluent Boston suburb of Newton. I'm here to meet a man who's been described to me as 'a firebrand' who 'doesn't suffer fools gladly' and 'leaves corpses everywhere'. We meet on the driveway. He's been in the yard preparing to lay some cobblestones. This is the sort of thing he likes to do on holiday (a holiday I'm interrupting, it turns out, adding to my nervousness). 'Sometimes I like problems I can solve completely alone,' he explains. Teamwork, it turns out, hasn't always come easily to him.

He's in obviously rude health – clear skin, piercing eyes, a frame that's clearly no stranger to exercise. At nearly fifty there isn't a hint of middle-aged spread about him and only the merest suggestion of grey in his short, dark brown hair.

'So, what's your deal?' he asks. 'Besides being an important writer that I have to meet?' I can't tell if that's generosity or sarcasm, because his features don't move much when he talks, as if anything overly expressive would amount to a waste of resources. His whole manner exudes ruthless efficiency. He'd make a great Hollywood villain. But there are flashes of charm too – and despite his reputation as something of a ball-breaker, at his core this is a man guided by a single, benign force. I don't think he could have achieved so much if he wasn't.

Jamie Heywood is a man driven by love.

In 1998 Jamie's younger brother Stephen, an architect and builder, found he couldn't turn a key in the door to one of the houses he was renovating. Soon after, the athletic and handsome Bostonian was diagnosed with Amyotrophic Lateral Sclerosis (ALS), more commonly known as Motor Neurone Disease in the UK and 'Lou Gehrig's disease' in the USA – a condition that erodes the nervous system's ability to control our muscles. Sufferers become progressively weaker over time, losing the ability to speak, swallow and, eventually, breathe. 'Luckier' patients (Stephen Hawking being the most famous example) may be spared fatal degeneration in the systems controlling the operation of their diaphragm and swallowing muscles, but they're outliers. For most it's a swiftly arriving death sentence.

When Stephen was diagnosed, Jamie immediately set about trying to save him, no small ambition given a) a cure for the condition had completely evaded the medical profession since it was first identified in 1824, b) Jamie, a graduate in mechanical engineering, had absolutely no medical training, and c) if Stephen's disease progressed at the rate of most sufferers he had less than four years. Within three days of Stephen's diagnosis Jamie had quit his engineering job in San Diego, relocated to the basement of the family home in Boston and incorporated the world's first not-for-profit biotech company with the sole aim of finding a cure. The first $10,000 to fund what became the ALS Therapy Development Institute (ALS TDI) came from Stephen. Jamie raised another $400,000 in the first year alone, and ten times that the following one – enough to rent premises (and refit them into what is now the largest ALS lab in the world) and attract leading researchers to his fledgling enterprise. Like I said; ruthless efficiency. And love. It's a hell of a combination.

Time was of the essence, which meant creating a new drug from scratch wasn't an option Even if ALS TDI discovered a new wonder drug *on day one*, it would probably be too late for Stephen; he'd die before the six years required to get it approved, a period dominated by ever more involved and expensive clinical trials required by the Federal Drug Administration.* (These trials rightly seek to validate any drug's effectiveness, determine ideal dosages and explore side effects.) Instead the strategy was to screen drugs already approved for the treatment of other conditions and see if they might *also* be effective against ALS. Doctors are allowed to prescribe drugs 'off label' (to treat a condition they weren't originally developed for) if there's good research to suggest this might help. In fact, drugs finding alternative uses to those they were originally developed for is common. For example, Raloxifene, now a breast cancer drug, was originally developed to treat osteoporosis. Sildenafil, initially proposed as a medicine for angina and hypertension, became one of the most lucrative drugs of all time thanks to its effect on an entirely different condition. It is now best known by its brand name: Viagra. If ALS TDI could find something that was already out there, the chances of saving Stephen were much higher.

---

* ... and that was being optimistic – an industry rule of thumb is that the average 'time-to-market' (from discovering a drug to general availability) is twelve years.

Stephen's disease progressed slower than average. Four years after diagnosis he was still alive, though chair-bound. Jamie describes his brother as 'invincible'. From his wheelchair (customised by Jamie) he continued to oversee house refurbishment projects, including the 'carriage house' next to Jamie's home (where he and I are now talking). Stephen got married to Wendy and had a son, Alexander ('equipped with his first, full-sized power drill at the age of two'). His sense of humour was legendary, emerging even in the prospect of death, insisting he wanted his end to be heroic – saving someone from a fire. But, he joked, it would have to be a fire that spread slowly, and there would need to be ramps because he'd be in a wheelchair.

In 2002 there was a hint that ALS TDI's strategy might pay off. A study from Johns Hopkins University School of Medicine showed that mice with a particular version of ALS lived longer if they were given the anti-inflammatory drug Celebrex. Here was a hot lead. 'Immediately Stephen's doctor and I collaborated and he started on the drug at a higher than normal dose,' says Jamie. At the same time he started trying to replicate the Johns Hopkins research because 'I'm an engineer by training, which means I like to validate things.'

The problem was they couldn't get the same result. In fact, in ALS TDI's rerun of the study there was no difference between un-medicated mice (the 'control group' in the parlance of scientists) and those that had been given Celebrex. Both groups died at the same rate. To be sure, they ran the study three more times with the same unhelpful result on each occasion: the mice dosed with Celebrex showed no advantage over the unmedicated 'control' group. How could this be? The conclusion Jamie reluctantly drew was that the original study was flawed. 'We realised there must have been something wrong with the control group in the Johns Hopkins study. Celebrex wasn't extending the lives of the mice that took it. Instead the mice in their control group had died *earlier* than average for some reason.'

But if that research was flawed, could other studies that ALS TDI had been basing some of its own efforts on be trusted? Luckily, from the outset Jamie had been intent on running his operation in a manner more akin to an engineering company than a traditional research lab: because they're often involved in building things that (if they go wrong) can instantly and spectacularly kill people (bridges, aircraft, roller-coasters, etc.), engineers place a strong emphasis on repeated and robust testing.

Accordingly, Jamie had created a lab whose mice numbers dwarfed previous experimenters in the field. By the time of the Celebrex study ALS TDI had already run trials involving over 10,000 mice with the disease, four times more animals than all other ALS studies in history combined – data it now subjected to a mathematical reality check called a Monte Carlo simulation. The results were horrifying.

Perhaps the simplest way to understand a Monte Carlo simulation is to consider the chances of getting a particular score when rolling a pair of dice. Listing all the possible combinations will soon reveal that you're more likely to roll a seven than any other number. Out of the 36 possible combinations, you'd quickly ascertain that six of them (or just under 17%) will yield you a '7', while the chances of getting a twelve (one combination) are just under 3%. A Monte Carlo simulation is a more complicated way to arrive at the same percentages, by rolling the dice many, many times (say 10,000) and noting down how often each number occurs. Over time you'd find that sevens occur roughly 17% of the time and twelves far less – and the more often you roll the dice, the more accurate the results become. While this technique is massive overkill for working out the chances of rolling a particular combination of dice, it is useful if you have a more complicated question to answer like, 'what's the likelihood of an unmedicated mouse with ALS living for 150 days as opposed to 170 days?' – and you have enough data to crunch.

ALS TDI's Monte Carlo simulation revealed a terrible truth: that at least half, and probably most, of the medicated mice in all previous ALS trials had lived longer as a result of random chance. To be sure they reran the actual experiments from a fifth of the previous studies (the most promising ones) in their own lab, but with much larger numbers of animals. Sean Scott, who led the research, told *Nature* 'we were heartbroken, because even using dramatically more animals than any of those other labs … we just could not get any of those drugs to work'. In other words, all of the previous studies into ALS were unwittingly bogus. Worse, clinical trials that had been set up based on those studies had been money and time down the drain. Unsurprisingly, in the light of ALS TDI's analysis, they all failed to replicate the false promise of the early studies.

The emotional impact on Jamie was enormous. He'd set out to save his brother, had found funding for, and built, the biggest ALS lab in the world – a lab he hoped would accelerate the pace of research. Instead he'd discovered that the field was, in reality, a long way behind where everyone thought it was. With time running out, Jamie was further away from his goal than ever.

How could the majority of medical research into ALS therapies be spurious and nobody realise? The answer is more disheartening that you might think, because it's not just ALS research that turns out to be suspect. Nearly the entire medical research profession suffers from bias and bogus results.

Much of the credit for this revelation goes to Dr John Ioannidis, who's built a formidable career in 'meta-research' (essentially research about research). His seminal 2005 paper 'Why Most Published Research Findings Are False' proved, scientifically, that many medical findings are based on shoddy research. His team at Stanford's Meta-Research Innovation Center continue to demonstrate over and over again that the conclusions of published studies in medicine (conclusions doctors collectively refer to when prescribing drugs) are often misleading, overstated, non-replicable or 'accurate measures of the prevailing bias'.

How can this be? The answer is that medical researchers, like the rest of us, hope. They hope their studies will yield results that answer the questions that bother them. Day to day, in order to keep that dream alive, they need to secure grant funding – which is much easier if their research is considered promising and published in the prestigious journals that the funders pay attention to. All of this can unconsciously guide their actions, says Ioannidis. 'At every step in the process, there is room to distort results, a way to make a stronger claim or to select what is going to be concluded. There is an intellectual conflict of interest that pressures researchers to find whatever it is that is most likely to get them funded.'

Ioannidis is clear that, while scientists may be knowledgeable about their specialisms, many are less able to design and operate a study that will put checks and balances on any unconscious bias. The potential potholes for an unskilled study designer are numerous. They pose the wrong questions, design studies without reference to existing evidence, recruit the wrong participants (or too few), take the wrong measurements or analyse data erroneously all within a system that encourages scientists to publish in well-regarded journals, who (to maintain their reputation) reject most of the papers submitted to them, and *especially* those that report negative results rather than positive ones. 'Currently we reward the wrong things,' says Ioannidis, 'people who submit grant proposals and publish papers that make extravagant claims', which means 'the hotter a scientific field … the less likely the research findings are to be true'.

For drugs companies, the hope takes a different form – the desire for greater profits. Take the case of Reboxetine, a drug for depression made by Pfizer. *The Economist* reported how the firm published trial data for the antidepressant showing a beneficial effect on over 65% of patients but neglected to publish the results of six further trials that, if taken into account, gave an average figure of just 11%. (Two of the unpublished trials actually showed patients fared *worse* on the drug.) As a doctor would you be more inclined to prescribe a drug reported effective 65% of the time, or 11%?

Harvard Medical School's Dr Marcia Angell, for two decades an editor of the prestigious *New England Journal of Medicine*, summed up the situation with extraordinary candour:

> 'It is simply no longer possible to believe much of the clinical research that is published, or to rely on the judgment of trusted physicians or authoritative medical guidelines.'

Since Ioannidis' initial research there has been a growing acceptance within medicine that there is a genuine problem to be addressed. An oft-quoted statistic is that dodgy research wastes over $100 billion a year. Depressing, isn't it?

I first heard Angell's quote as part of a speech given by Jamie Heywood at the Drug Information Association's (DIA) 50th Anniversary conference – a talk he gave after being awarded the President's Award for Outstanding Achievement in World Health. The DIA is an industry body that fosters cooperation between those working in drug development and their colleagues in medical communications. Given how vocal Jamie's been about what's wrong with both industries, his selection raised a few eyebrows. But it also indicates how far he's come. Today Jamie, an engineer without a single medical qualification, is seen as one of the most important thinkers in healthcare.

In his talk he told the dramatic story of ALS TDI's damning findings, how this became 'a big scathing scandal' written up in *Nature* and how, 'as usual in medicine, nothing changed ... The evidence is that we're selectively distributing data – and you guys all know this'. And he didn't stop there, going on to quote research showing that preventable medical errors, once shockingly the sixth leading cause of death in the United States, have now, unbelievably, crept up the table to third place. 'Hospitals make huge amounts of money in their Intensive Care Units, where they carefully put people, by giving them pneumonia. My brother used to go to the hospital with an illness and they'd give him *another one* and pocket $30,000!'

'So, the question', Jamie asked his audience, 'is how do we do it better?'

Clearly not by doing more of the same.

Despite Jamie's best efforts Stephen died in 2006, when a respirator supporting his weakening diaphragm became accidentally detached the day after Thanksgiving. Amazingly, after forty minutes of CPR, Stephen's heart restarted but he was brain dead. His body remained alive long enough to be harvested for organ donation. He was thirty-seven.

ALS remains incurable. Tens of millions of dollars and sixteen years since its formation, ALS TDI has, at the time of writing, failed to find any drug with an impact on the disease. In fact, the company's biggest contribution to the field has been to undermine previous assumptions that the disease starts in the nervous system. (Their research suggests that the earliest physical change actually occurs in the 'neuromuscular junction' – where your nerves enter your muscles.)

The journey had also cost Jamie his marriage. 'I'm not the easiest person to be around,' he says when we touch on the strains that Stephen's diagnosis, care and the search for a cure put on the family. And you'd forgive Jamie if, at this point, he'd elected to end his foray into medicine. But he'd been given four things by his brother that made this impossible.

First there was the anger. 'Drug discovery is like a broken religion,' he says. 'It's full of priests who think they serve God, but they really serve themselves and they're seductive and they're powerful and they use language to confuse people.' As he says this you can hear the rage in his voice, but not because he raises it or speaks more quickly. The irritation is deep, it's in the tone, an anger that's been tamed, burnished, made useful.

Second, Stephen had given Jamie a powerful voice and perspective. He became the archetypal 'guerrilla scientist'. His brother's predicament gave Jamie a moral authority to confront the status quo, to question the system in a way those acculturated to it could not. 'I was able to see where the power lines were, what controlled people's behaviour, the assumptions not questioned … I was able to challenge things.'

Then, seemingly out of nowhere, Jamie says one of the most extraordinary things I've ever heard. 'You know, I think I would be a miserable, ordinary person if Stephen hadn't got sick.' It stops me in my tracks. Declaring 'my current happiness stems from the illness that killed my brother' is a hell of statement. But Jamie's not trying to shock, he's just being searingly honest. 'I mean, I am deliriously happy,' he continues. 'I work on exactly what I want to work on. And the friends I've made? Such amazing people I would never have met otherwise.'

Happy he may be, but this does not translate into easy laughter or an impression of contentment, however. One of his favourite observations is that of choreographer Martha Graham who spoke of the joy that comes from 'a queer, divine dissatisfaction' and the 'blessed unrest that keeps us marching'. The third thing Jamie got from Stephen was a life he couldn't have imagined for himself, a life of purpose, of useful anger, a drive he never had, an itch he has to scratch.

'Is that bittersweet?' I ask. 'If you could go back in time and somehow stop Stephen from getting ill ...?'

'Stephen and I talked about that. I'm not sure anyone would go back and undo it. Stephen most of all.' He pauses. 'If you go back and regret, you have to say that you don't like where or who you are, and if you don't like who you are that's really sad.'

Everyone told me Jamie Heywood would be a difficult interviewee, that he'd test me and I'd be unlikely to warm to him. I'm not feeling it. Yes, he's exact, yes he's angry, yes he's driven, and for sure he's never going to have a career as a stand-up, but it's hard not to feel well disposed to a guy who, faced with tragedy, chose action over acceptance and then, even when Stephen died, carried on challenging the system that failed him and his family. Despite all the warnings I like the guy. Not that he'd care, of course. 'You know, I just want the next Jamie to enter a system that has some *logic* about it,' he says.

It's this ambition that's brought me to Boston, to talk about Jamie's solution to many of healthcare's woes. The catalyst for his innovation, once again, was love.

2 PATIENTS LIKE ME

'The company in which you will improve most, will be the least expensive to you.'

– GEORGE WASHINGTON, FIRST PRESIDENT OF THE UNITED STATES

In January 2007 Dave deBronkart went for an X-ray to investigate a pain in his shoulder and got considerably more than he bargained for. His doctor spotted a shadow on the right lung and ordered further scans that revealed, as deBronkart puts it, 'I was already almost dead'. Not only did he have kidney cancer but that cancer had spread ('metastasised' in the parlance of oncologists). There were tumours in his lungs, muscles, his tongue and several bones (which, as the cancer advanced, caused his leg to break one day when he put too much weight on it). His most likely prognosis: about six (not very pleasant) months to live. 'This was bad,' he remarks with superhuman understatement.

DeBronkart's doctor recommended he join acor.org – the Association of Cancer Online Resources – a network of cancer patients using rudimentary web forums and email to share their experiences. They confirmed his doctor's diagnosis: the chance of a cure was almost impossibly small. But, it turned out, not zero.

His fellow sufferers told him of a long shot: a high-dosage of the drug Interleukin, which had shown dramatic results in a small subset of patients. That 'small subset' is a problem for some physicians, who are sceptical about whether those documented successes are statistically relevant. Coupled with the knowledge that Interleukin comes with some pretty unsavoury side effects (including fever, vomiting, diarrhoea, confusion, drowsiness and memory loss), many kidney cancer specialists are doubtful about the treatment. In fact, they won't even tell you about it.

This is a classic example of patient motivations coming into conflict with those of the healthcare profession. Interleukin's ability to save a very small number of kidney cancer sufferers hardly touches the average survival rate. Statistically speaking, it's a non-treatment. But what if you're one of the lucky 'good responders'? The fact your experience doesn't translate into 'statistical significance' is hardly the point. As far as you're concerned there is only one statistic that matters. You: 1, Grim Reaper: 0.

Luckily for deBronkart his new online friends were able to give him the phone numbers of four doctors who saw it the same way. A regime of Interleukin was a tough ride for sure, but the cancer made a dramatic retreat … deBronkart had beaten the disease.

The experience gave Dave deBronkart a new calling, as an evangelist for the patient's voice in re-imagining healthcare. He co-founded and became spokesman for the Society for Participatory Medicine – largely inspired by one of his heroes, Dr Tom Ferguson, who back in 1996, before anyone had even heard of Google or Wikipedia, published the first book on how patients could use the web to improve their outcomes.* 'You are already your own doctor,' he wrote, quoting research showing 'people provide their own illness care between 80% and 98% of

---

* It's called *Health Online: How to Find Health Information, Support Groups, and Self-Help Communities in Cyberspace.*

the time.' He argued that 'self-care is, and has always been, our predominant form of healthcare'. This wasn't just an academic point for Tom, it was his personal reality. He managed his own multiple myeloma (a blood cancer) for fifteen years (far exceeding the average life expectancy of four years). Shortly before the disease finally took him in 2006, he wrote that the world needed:

> 'a new cultural operating system for healthcare in which e-patients can be recognised as a valuable new type of renewable resource, managing much of their own care, providing care for others, helping professionals improve the quality of their services, and participating in entirely new kinds of clinician-patient collaborations, patient-initiated research, and self-managed care.'

Building this new 'cultural operating system' would, he argued, be the only way to fix 'the seemingly intractable problems that now plague all modern healthcare systems.' In other words, rather than the healthcare system needing to improve itself so it could better serve patients, patients themselves were the key to fixing things.

It's 2005 and Jamie Heywood, following the break-up of his marriage, is looking for romance. Like many singles he's turned to the Internet, browsing a dating site.

Successful dating sites need two things: lots of members and lots of data about them. Together these two things increase the chances of making potentially compatible matches. Hopeful singles are provided with a wealth of search criteria, while the sites' algorithms attempt to match people who, based on their data and preferences, might hit it off. The more information members share, the better the likelihood of a good match, which is why online romance hunters are often happy to reveal a surprising amount of personal information – their height, weight, hair and eye colour, where they live, their level of education and romantic history, hobbies and interests, their job and an indication of how much they earn, along with their own preferences in a mate – all stored in a searchable format that's amenable to interrogations from each site's matching engine. A computer systems geek would say the data is 'computable'.

As Jamie browsed he wondered why there wasn't a matching system for patients, where they could find sufferers similar to themselves – the same sex, same age, at the same stage of any disease's progression. Perhaps somewhere they could share information about medications (and dosages) and how their symptoms fluctuated in response to different treatments? Could Stephen have benefited from connecting and comparing experiences with fellow ALS sufferers of his age, build, sex, at the same stage of the disease? After all, Stephen had a lot of data to share. Jamie's singular focus and data-heavy engineering approach to finding a cure meant his brother was probably the most documented ALS patient on the planet.

Everything had been recorded. Not just the dosages of various medicines and their effects, but also the severity of all his symptoms over time, from constipation (a real problem for ALS sufferers) to excess saliva, to the frequency of involuntary muscle twitches. Jamie's family had the results of every medical test Stephen had undergone, his 'functional rating score' (a scale for assessing the physical impact of the disease) over time, his weight and a record of his subjective experience, how he felt. (As a family it had been a just as important to understand what helped Stephen be happy as it was to track the physical symptoms of the disease.)

This wealth of data was Stephen's fourth gift to Jamie; inspiring him to imagine a platform where patients' information was as computable as that on the dating site he was looking at. If it could work, surely this would be part of the 'new cultural operating system for healthcare' that Tom Ferguson had imagined – a place for patients to improve their self-care through sharing knowledge with each other, but also providing a hub of data that might offer insights for the medical profession trying to develop new treatments. After all, if the analysis of aggregated online dating information tells us that (no word of a lie) vegetarians are keener on oral sex than non-vegetarians, that atheists have better spelling and grammar than the religious, or beer drinkers are more likely to put out on a first date, what surprising and hitherto unknown insights might come from a successful patient-matching site?

Jamie immediately set to work with his other brother, Ben, and a family friend (and veteran of the dotcom boom), Jeff Cole. Poring over Stephen's data, they began to ask how other sufferers of the disease could most easily record the same amount of information, in the same detail, and how that data could be shared and interrogated. The result was a website for ALS sufferers called PatientsLikeMe.

The critics were quick off the mark. Patient-provided data would be poor quality, they said – anecdotal and not suitable for rigorous analysis. That was if you could get patients to regularly submit their data in the first place. And, even if they did, would they really be willing to share it so freely? The privacy of our medical records is one of the pillars of a modern healthcare system; our health is a deeply personal issue. Would ALS patients really want to record and then broadcast their daily battles with constipation? On top of that, such a network could put patients at risk. The online world is already awash with quackery and snake oil, and patients without the training of a physician could easily misinterpret data or spread dangerous or misleading information to each other.

Jamie had a hunch patients would take a different view. It's well documented that slimmers who keep a food diary lose twice as much weight as those who don't. The act of recording what we eat raises our self-awareness, so that a bag of crisps snaffled absent-mindedly with your morning coffee is brought into the light and questioned, rather than remaining a quickly forgotten and habitual calorie bomb. Provided with an easy mechanism to record – for instance, the drugs they were taking, in what dose and when – patients might also begin to gain useful insights into managing their condition; maybe noticing they felt better if they took certain drugs with food, or at particular times of day.

On the matter of sharing their data with each other (and the world), Jamie figured that for sufferers with diseases like ALS confidentiality would be a long way down their list of worries. If there was a chance that recording and sharing their data could help them improve their day-to-day experience (and if the aggregated data from the site might accelerate the search for a cure), then pragmatism would outweigh privacy concerns. Jamie also had a sneaking suspicion that some objections to his idea were based not in concern for patient care but because, if it worked, it would challenge the position of physicians as the sole authority on how you should manage your condition.

Finally, the idea that data recorded by patients might be inaccurate was a bit rich coming from the medical profession. The studies can't agree how bad it is, but they all admit that errors in our medical records are common. Frequent errors include mistaken diagnoses being logged, medicines the patient is taking not being listed, incorrect dosages recorded, the patient's reported symptoms remaining undocumented, inaccurate treatment outcomes being noted down and lab results going missing. The reasons for such errors are numerous – everything from ticking the wrong box on a list, to poor handwriting, to prejudice (some studies suggest that patients who are perceived difficult or come from a racial minority suffer more errors on their records than the statistical average).

And then there's the big banana of cock-ups: information getting recorded on the wrong patient's file, which, if you read around the literature, you discover happens far more often than you might hope. Take the case of Mary Kerswell, who, in 2012 was called in to her local surgery in Biggleswade, Bedfordshire, to take an urgent test related to her kidney condition. Except Mary didn't have a kidney condition. (Apart from a touch of glaucoma and being a little hard of hearing, the sixty-seven-year-old retired scientist was in rude health). Being a curious and determined sort she asked to see her medical records. When these didn't materialise at the agreed time (and after Mary had paid a fee to see them), she performed a sit-in at her local surgery. The police were called, she was hand-cuffed and removed from the scene. When she eventually did clasp eyes on her data Mary found out that, not only did she (apparently) have kidney disease, she was also a heavy smoker, suffering from Alzheimer's disease and had undergone both a hysterectomy and a double hip replacement. None of this was true. Somewhere along the line Mary's records had become mixed up with another patient.

What's worse is that the current drive to create electronic health records is, it seems, exacerbating the problem. As data is translated from reams of paper into computer files, extra errors (on top of those already in our files) are being added in. Our data is 'lost or incorrectly entered, displayed, or transmitted.' How about, for instance, the voice recognition dictation software that reportedly heard 'DKA' (shorthand for 'diabetic ketoacidosis') as 'BKA' ('below the knee amputation') and entered it on a patient's new e-record accordingly. (Worryingly, it took four subsequent hospital visits for the error to be noticed, despite the fact the patient was, I'm assuming, happily *walking* into those appointments.)

Dave deBronkart's cry of 'Give me my damn data, because you people can't be trusted to keep it clean' is hard to argue with. Thanks to campaigners like him, there are increasing calls for patients to have easy access to their own medical records. When they do get them, the results are promising, with errors spotted and fixed. So, whilst the introduction of electronic patient records clearly risks *increasing* errors, it also offers a chance to reduce those same errors (and add in useful detail) in the long term, *if* they can be seen as a joint document created and overseen by both patient and healthcare professionals.

Jamie's vision is to expand the data recorded further, bringing patient and physician to a closer, joint understanding of the patient's condition. 'We need the data recorded by healthcare professionals, for sure,' he says. 'It's vital. But we absolutely need patient-provided data as well. It's the only way you can get a full picture. And if you have a whole bunch of other patients' data stored in structured format, in a *computable* format, you can begin to change the way medicine is done.'

Patientslikeme.com was launched for ALS patients in 2006 and no one outside that community really noticed. But Jamie's intuition proved well-founded. Patients began to use the tools provided to record their symptoms and treatments and find patients like themselves. (I look at the site and see over 8,000 ALS patients have registered since the launch, 65 in the last month.)

Then in 2008, with Jamie's help, those patients did something extraordinary. 'So this paper comes out that suggests lithium carbonate can slow the progression of ALS,' Jamie recalls. 'To be honest it's a pretty crappy trial with a small number of patients and controls, but it shows this positive effect and so it's big news in the ALS community.'

Only sixteen patients received lithium carbonate in the study, so whilst the results were intriguing they certainly weren't conclusive. They were interesting enough however to stimulate the National Institute of Neurological Disorders and Stroke (NINDS) to initiate a large 'randomized, double-blind, placebo controlled trial of 250 subjects'.

This is the kind of trial that meta-researcher John Ioannidis has found least susceptible to bias. When large numbers of patients take part, statistical curios and blips are hopefully ironed out. A 'randomised' trial is one where patients participating are selected at random (usually by computer) to receive the drug under test or a placebo (a sugar pill). This helps reduce any unconscious bias on the part of researchers (e.g. giving the drug to the 'healthier' patients at the outset – which could favourably skew the results).

The 'double-blind placebo' trial is a further check and balance, in which the computer's choices as to who gets what is hidden from the researchers as well. This 'double-blinding' eradicates the risk that researchers might inadvertently give clues to participants as to what they are getting, guarding against the strange phenomenon that is the placebo effect – where, if a person *expects* a treatment to have a certain effect, their body's own chemistry can sometimes produce similar results to those the medication might have caused.* This means any therapeutic effects observed are far more likely to be down to the drug alone and not influenced by a patient's in-built placebo response.

Most studies aren't this well populated or controlled, for the simple reason that large, randomised, double-blind, placebo-controlled trials are mightily expensive. Recruiting patients takes time. Administering all those checks and balances costs money. If patients are spread out geographically (often the case for trials dealing with rare diseases like ALS), then several research centres have to be involved (which adds further administrative expense), and the longer the trial runs, the bigger the bill. An evaluation of 28 such trials run by NINDS revealed an average bill of $12 million, – which is one of the reasons why what happened on patientslikeme.com made people sit up and take notice.

Even before the initial study showing the possible benefit of lithium carbonate was published, news of its hopeful results had spread in the ALS community on PatientsLikeMe. Ahead of publication the study's authors had presented their findings at an Italian-language conference, and resourceful ALS sufferers in the USA used Google Translate to extract the key findings. Lithium carbonate was already available as a treatment for bipolar disorder, and within two months more than 160 ALS patients on

---

*   These effects can be positive or negative. If a patient believes they may suffer unpalatable side effects, these can manifest too – the so called 'nocebo effect'. Nobody knows how our bodies do this (it's one of the most enduring conundrums in medicine) but it's the reason, for instance, homeopathy 'works' for some people. In fact, it turns out that the more convincing and involved the story about a placebo, and the more time that is spent administering it, the more our bodies respond.

PatientsLikeMe had taken a calculated risk and started taking it off-label, collating their experiences (different dosages, side effects etc.) on a shared Google spreadsheet. A medical researcher reading this will almost certainly have alarm bells ringing. Official medical trials are centred around 'informed consent', meaning patients must understand and be knowledgeable about the risks they're taking before they're allowed to take part. Could ALS patients on PatientsLikeMe, eager for new treatments, have their judgement clouded by their hope – ignoring the risks? Almost certainly. If scientists can be swayed in their study designs by unconscious biases, it would be ridiculous to suggest that patients hoping for a possible postponement of their death wouldn't be. But that also meant, informed consent or not, they were going to do it anyway.

Jamie and his team realised that there was an opportunity to create a new kind of drugs trial: 'We quickly built a lithium-specific data collection tool for them and also a matching algorithm pairing those experimenting with lithium to non-experimenters.' Because patients on the site had been regularly recording their 'functional rating score', those taking the lithium could easily be paired with virtual 'controls' whose disease progression had previously followed a similar pattern. (In fact, each experimenter was paired with *three* controls.) The two groups were then compared going forward in a unique 'virtual' trial. It wasn't 'randomised' (the patients who took the drug had actively elected to do so) and for the control group it wasn't really a trial at all; they weren't active participants getting a placebo but users continuing to record their day-to-day data on the site as they always had. Nonetheless, for the minimal cost of building a new data collection tool and a matching algorithm, PatientsLikeMe had a drug trial on its hands – a trial that had 'recruited' its participants almost instantaneously and in larger numbers than the proposed NINDS study.

Before the NINDS trial had given its first dose of lithium carbonate or placebo, PatientsLikeMe was able to predict what the results would be. (The NINDS study confirmed PatientsLikeMe's analysis a year later.) The bad news was that lithium carbonate turned out to be useless in the fight against ALS. The good news was that by using a body of accessible day-to-day 'structured, computable health data' PatientsLikeMe had beaten the 'gold standard' of clinical research to its conclusion at a fraction of the cost. It's perhaps the company's most famous success – and the one that had the business press soon heralding it one of 'fifteen companies that will change the world' – but it's not the only one.

Soon after launch PatientsLikeMe expanded beyond the ALS community to cover other conditions. (If you visit patientslikeme. com today you'll find data-recording and sharing tools for nearly a hundred separate ailments, and over 440,000 patients registered.) As those new communities grew, so fresh insights emerged. From a 2009 discovery that shows compulsive gambling in Parkinson's sufferers (a side effect of the medication in a subset of patients) is likely to be twice as prevalent than previously thought, to a 2014 finding establishing a link between the severity of multiple sclerosis symptoms and the onset of menopause, PatientsLikeMe has started to prove that a patient network built around structured data has a lot to contribute.

However, one study catches my eye above all others: a simple survey of the epilepsy community conducted in 2011. Asked about their experience since joining the site, a third felt they now received improved care thanks to a better informed relationship with their physicians; 27% said joining PatientsLikeMe had helped reduce the side effects of their treatment, while 18% claimed fewer visits to the emergency room since logging on for the first time. As Jamie's brother and co-developer Ben remarks, 'If I could create a drug that could do that I'd be a very rich man.' (For two patients there was another unexpected and life-changing side effect to joining the community. They got engaged; a heart-warming echo of Jamie's original inspiration.)

It's with good reason that patient engagement has been called the 'Blockbuster Drug of the Century.' Research shows that patients who are actively involved in their own care spend less time in hospital, manage their conditions better, are subject to fewer medical errors, are more likely to engage productively with their healthcare providers and will have a higher opinion of them. Unsurprisingly, they also place less of a financial burden on the healthcare system – about 17% less, according to Dr Judith Hibbard, Senior Researcher at the University of Oregon's Health Policy Research Group. As Dave deBronkart is fond of saying, 'The most underutilised resource in all of healthcare *is the patient.*'

There's a reason drug companies like Merck, Genentech, Actelion, Biogen and AstraZeneca have entered into partnerships with PatientsLikeMe. They've realised Jamie's once-mocked idea could help them develop products that work better for patients in the real world. And, as I will soon find out, drugs companies are in desperate need of help. There's another interesting partnership, too, with the US Federal Drug Administration – exploring ways patient-provided data can deliver new insights into how already approved drugs are panning out 'in the wild'.

As endorsements of Jamie's idea these deals are hard to argue with. But it's the personal stories of patients helped by the site that are his real reward – patients like Letitia Browne-James, who I'd talked to the week before flying to Boston. Suffering since childhood with 'refractory epilepsy' (a catch-all term for epilepsies that respond poorly to drugs or confound brain scans) and enduring up to sixteen seizures a day ('some so severe that I would bite my tongue and cheeks so badly I couldn't eat properly for a week because of the pain'), her epilepsy was like a poltergeist, detectable only by the havoc it wreaked, an unseeable and seemingly untreatable destructive force, hiding somewhere in her brain. She couldn't drive and could barely hold down a job. She joined the site 'out of sheer desperation' and after convening with her fellow sufferers for a few months (the site has 10,000 registered epileptics) noticed several had seen an 'epileptologist' (a specialist in her condition that her neurologist, she told me angrily, hadn't seen fit to mention existed). She immediately asked for a referral and after a barrage of tests overseen by her new doctor was declared a perfect candidate for a 'left temporal lobectomy'.

'I was like, "great!" ' she told me, laughing. 'My husband was more, "Babe, calm down! They're talking about cutting in your brain!" ' But her epilepsy was so debilitating she 'just knew that there was nowhere to go from where I was at but up'. She'd also checked out the success rate for others on PatientsLikeMe: 83%. On her profile, you can see the seizures she was recording daily stop dead the day of her surgery: 16 August 2012. Letitia's poltergeist had been evicted. 'I stayed stuck for thirty years because I didn't have the resources I finally got through that community,' she said. 'What it did for me, you can't imagine.'

Today she's flourishing, a clinical manager in the healthcare sector and pursuing a doctorate in Counsellor Education. 'We've just asked Letitia to join our Patient Advisory Committee,' says Jamie. 'And we filmed her telling her story.' (You can watch the results on YouTube). 'There's a moment that she puts the key in the ignition of her car that kills me.' For a brief moment Jamie's face loses some of its calm efficiency.

Then there's Jackie, who suffers (along with 50,000 other registered users on the site) with multiple sclerosis. 'I remember trying to Google for information and it was always insufficient,' she says. 'When you find a site like Patientslikeme ... you realise that there are literally tens of thousands of people that share your condition and your struggle and your everyday battles.' One of those battles was dealing with her medication, which was actually making things worse. Thanks to the patient data on the site she found an alternative and 'took it to my doctor and I'm having very good luck with that medication'.

'PatientsLikeMe might not succeed in the long run,' says Jamie. 'Someone could come along and do what we do better.' That however is not really the point. The idea that patients should be part of a new 'operating system for healthcare' is out there and it's not going away. Jamie's proved the idea can, and does, work at scale. Not bad for a guy with no medical training. He set out to find a drug, and failed. Instead he found something more powerful than any drug ever created – a way for patients to improve healthcare. 'You must be pleased,' I say as we walk back out to the driveway – Jamie wants to get back to those cobblestones and it's a beautiful sunny day, a good day to be outside.

'Look, I'm an engineer and for an engineer there is only one question: "Does it work?" In science, in medical research, the question is too often "Do I *own* the idea, do I get published, where's the credit, can I make some kind of profit out it?" I just want to build a platform where we have a better chance of saying to patients "given everything that's out there, given what everyone else is trying, the best possible outcome for *you* is *this* – and here's how to get there."'

'I think that's a "yes",' I say.

He replies as you might expect an engineer to, describing the mechanics rather than the experience of his emotions.

'I feel like I'm doing well … but I have this image of a shark. You know the myth that if they stop swimming they die?'

I nod.

'Well, I've set up an environment where I can do an awful lot of swimming.'

A few days later I'm sitting in a Thai restaurant in downtown Boston with one of my favourite thinkers, Juan Enriquez. I don't always agree with him, but as a man who is variously an investor, futurologist, former peace negotiator, author, Harvard academic and businessman, he is never dull company. Juan's obsessed with the future and how to make it better, and we're meeting shortly after he's finished his latest book, *Evolving Ourselves* (co-authored with scientist, investor and entrepreneur Steve Gullans), exploring the next steps in human evolution. He gets around a lot – and I want his opinion on where I might go in my search for systems-level innovations.

'If you want to see innovation, real innovation, the best advice I can give you is go to the places where things are most broken,' he says. 'Go where the existing system has failed people really badly, or there's no system at all – and that's where you'll find the interesting stuff.'

I tell him about a few stops I've already got planned. 'So,' he says, 'you already figured this out, then.'

3 BUG IN THE SYSTEM

'Sir, an equation has no meaning for me unless it expresses a thought of God.'

– SRINIVASA RAMANUJAN, MATHEMATICIAN

Jamie Heywood failed to locate the magic drug for his brother, but his story has inspired me to delve deeper into the world of finding new medicines. Because right now the system is expensive, inefficient, corrupt and essentially immoral. It condemns millions to die an unnecessary death. And so I've come to Delhi to meet a team building an alternative, a team led by the wonderfully jolly Samir Brahmachari.

Samir is not your typical scientist. By his own admission he's 'crazy and emotional' and, although well into his sixties, this rotund and ebullient man somehow conveys the demeanour of a young boy who's just got away with something deliciously naughty. 'Great science, great literature, all great philosophy comes out of *emotional* engagement! Not by being professional!' he exclaims. 'That's why I don't like people who will say "Yes, yes, yes" to all of everything! None of the people you've come to see were picked because of their performance in examinations. No! I look for people who can question me, people who can challenge my authority!'

He turns to a young woman next to us, Dr Anshu Bhardwaj, a 'project investigator' at Delhi's Institute of Genomics and Integrative Biology (the building in which we're all sat).

'Tell him how we met!'

Anshu's open, intelligent features break into a smile at the memory. 'I was at a conference in Bangalore, presenting some of my research,' she recalls. 'I really wanted to get his comments, because that would be awesome.' (Despite his unconventional approach, Samir is pretty much India's highest-ranking scientist.)* 'So, I was running after him for two days until I really ended up shouting at him, saying, "If you want to come, come right now! I can't run after you!" ' That caught Samir's attention. Ten minutes later, and after a few searching questions about her work, Anshu had a job offer, recruited to the project I've come to visit – a project that aims to redefine how we find new drugs, and save millions of lives in the process.

As Anshu discovered, if you're not prepared to challenge Samir, you must be prepared to be ignored by him. And if you *are* prepared to challenge him, you must be prepared to work extremely hard. As much as he respects the bravery to question the status quo, he also values the persistence required to come up with something better.

Despite the demands placed on them, the staff here don't hide their affection for the professor, even if sometimes their words are accompanied by the occasional rolling of the eyes. Nisha Chandra, another of Brahmachari's recruits, who'd generously picked me up from my hotel this morning (a kindness I much appreciate given the exquisite madness of Delhi's traffic), told me how she found him 'completely inspirational from the first day'. She was particularly taken with the non-hierarchical way Samir

---

* Until recently Samir was the Director General of India's Council for Scientific & Industrial Research (CSIR), the nation's research behemoth with nearly forty laboratories under its jurisdiction. It was a job that proved a special kind of hell for him – the maverick asked to manage a long-established and august national institution with all its attendant bureaucracy. Having stepped down from that role, he's happy being 'just a Professor again'.

runs his lab, something she makes clear is atypical in Indian academia and society more widely. 'He's the biggest person in the country for science and technology, I mean *seriously*, and to be in the same room and him talking to you one-to-one is just unreal. You wouldn't see that anywhere.'

Professor Brahmachari could be enjoying his retirement. Instead, most days you'll find him here, usually late into the evening. There's a reason he's so driven. He's trying to beat one of the biggest threats facing humanity, one that threatens to kill billions. And time is running out.

In 2006 a new kind of tuberculosis (TB) made an appearance. Dubbed 'totally drug-resistant', it is incurable. If it becomes active in your system you're almost certainly looking at a death sentence.* What's worrying is that the arrival of this new agent of death didn't really come as a surprise. Tuberculosis has been evolving its way around our drugs for over forty years. Worldwide, tuberculosis kills about 4,000 people a day, and one person a minute here in India. According to the World Health Organisation (WHO), it's 'second only to HIV/AIDS as the greatest killer worldwide due to a single infectious agent'.

So you'd think drugs companies would be beavering away researching new TB medicines, wouldn't you? But they're not, and they haven't. In fact, there hasn't been a new front-line drug for TB since the 1970s. Why? Simply put, because TB is

---

* If you have the commonest form of the disease it's still treatable, but this isn't a simple injection or a few pills – rather a six-month course of multiple anti-bacterial drugs. Hardier strains can also be treated, but patients are faced with two years of pill popping, taking a cocktail of 'first line' and 'second line' medication with some pretty unsavoury side effects. On these longer regimes many stop their treatment, feeling the cure is worse than the disease. They die.

a disease of the poor, and the poor can't pay enough to cover the multi-billion-dollar costs that drug companies say it takes to develop new medicines. Millions die because there isn't a drug company profit in them staying alive.

And it's not just tuberculosis. All sorts of diseases are becoming drug-resistant thanks to a massive decline in new antibiotics coming to market since 1980 – the period between then and now dubbed a 'discovery void'. Newly resilient forms of typhoid, malaria, influenza and E. coli (responsible for a host of different conditions, including diarrhoea, urinary tract infections, respiratory illness and pneumonia) are of particular concern. Meanwhile pharmaceutical companies have concentrated their efforts on developing treatments for so-called 'global diseases' (ones the rich get, too) like cancer, or 'lifestyle conditions' (where the rich will pay) like hair loss. Everyone is hoping for the next 'blockbuster drug' (one that makes huge profits), explaining why, for instance, there's a flurry of activity at the moment around developing drugs for obesity – a market predicted to grow 40% every year until the end of the decade.

But what's a pharmaceutical executive to do? A popular statistic is that for every 5,000 potential medicines that enter their drug development process just *one* is eventually approved for human use. (Even those that get as far as human trials have a less than one in four chance of making the final journey to prescription.) Drugs companies argue, you might think reasonably, that the expense of researching potential drugs that never see the light of day *has* to be covered in the price of those that *do* reach the market; otherwise they'd go out of business and there wouldn't be any new drugs. The 5,000:1 statistic has almost become a badge of honour for them – 'see how hard it is?' – and the reason the industry-funded Tufts Center for the Study of Drug Development puts the cost of bringing a drug to market at $2.6 billion.

Which sounds like a fair if not necessarily cheery argument. It's an argument, however, that fails to chime with a March 2012 analysis of the industry published in *Nature*. Entitled 'Diagnosing the Decline in Pharmaceutical R&D Efficiency', this looked at the number of new drugs approved per $1 billion spent. The authors concluded it's been halving 'approximately every nine years since 1950'. Or, in simple terms, the commercial drug industry has suffered a consistent and precipitous decline in productivity. With a touch of wry humour, the authors dubbed this phenomenon Eroom's law, the reverse spelling of Moore's Law – the famous law of computing that expresses the doubling of processing power per dollar every two years (meaning an average smartphone now contains as much computing muscle as was used by the entire Apollo space program).

Why this is happening is a hotly contested topic with everybody blaming everyone else. It's the growth of draconian, needlessly complex or over-cautious legislation that creates billion dollar hoops that drugs companies must leap through. (As Jamie Heywood from PatientsLikeMe told me: 'We're approving between 30 and 50 drugs a year when there are *7,000* diseases. We need to re-invent the whole system.') Or it's because drug companies don't share their knowledge, adding unnecessary duplication and cost into the industry, 'wasting resources and careers' as Chas Bountra, Professor of Translational Medicine at Oxford University, puts it. Or is it because drug discovery has had its early wins and is now having to target ever harder (and therefore more expensive) problems?

All of these arguments hold some water but there's another cause, clearly outlined by the authors of *Diagnosing the Decline*. Previous good returns from the blockbuster drugs of yesteryear have had a ruinous effect on the industry's ability to innovate. For example, Pfizer's anti-cholesterol pill Lipitor (reportedly the second-best-selling product of all time, after the Sony PlayStation) generated almost inconceivable revenues of $141 billion. With that kind of cash coming in, the big drug companies have historically adopted a 'throw money at it' tendency, say the creators of Eroom's law, increasing budgets and staff to create larger, better-equipped but no more efficient operations. As one very senior pharma exec said to me, 'The problem with our business is that success has bred mediocrity.'

Rohit Malpani, Director of Policy and Analysis at Médecins Sans Frontières (MSF) agrees. His response to the Tufts $2.6 billion figure was unforgiving. 'If you believe that, you probably also believe the earth is flat,' he said, pointing to research from the Drugs for Neglected Diseases Initiative (co-founded by MSF), which suggests it's possible to create new drugs for less than $190 million. MSF (and others) argue that the commercial drug industry is caught in a vicious circle of self-delusion – and it's costing millions of lives. Worse, factories used by the big pharma companies in China are accused of discharging antibiotic compounds into the surrounding environment, creating breeding grounds for drug-resistant 'superbugs'.

Large pharma companies are now more about marketing than drug-making, spending more on convincing doctors to prescribe their existing products than they do trying to create new ones – and not always in the most ethical fashion. According to an investigation by the BBC, until recently paying bribes to doctors 'was commonplace at big pharmas, although the practice is now generally frowned upon and illegal in many places'. Frowned upon, but the temptation is there. Of course, no one calls them 'bribes' – instead they're 'payments for promotional talks', consulting fees, free travel or (mostly) nice meals out. And it works. Researchers from the University of California discovered that doctors who get treats from pharma companies were twice as likely to prescribe their drugs. This is the 'respectable' end of pharma-to-physician payments, but corruption is also in their repertoire.* Let's take GlaxoSmithKline. In late 2014 the company received a $490 million fine after being found guilty of full-on, no-bones-about-it, bribery in China. Former head of Chinese operations Mark Reilly was given a suspended three-year prison sentence and deported. This is the same company that pleaded guilty to promoting the use of antidepressants to patient groups they weren't approved for (notably under-18s) and was landed with a mammoth $3 billion fine as a result.

Samir has his own depressing story. 'Initially I worked in the sincere belief that drug companies were the right vehicle.' His watershed moment came after he developed a test that could tell if a patient would benefit from an expensive asthma drug or not. 'The company didn't want it because if the test came back negative they would lose the revenue. "Give us something that will keep the patient paying," they told me.'

I could fill this entire book with examples of shoddy practice from all the major pharmaceutical companies. I'll save the paper (and direct you to Ben Goldacre's *Bad Pharma* instead), but you get the picture. Meanwhile drug-resistant bacteria are queuing up in their trillions to kill us, and not just in the 'developing' world. According to the US Centers for Disease Control and Prevention, 'Each year at least 2 million Americans become infected with bacteria resistant to antibiotics, and at least 23,000 die.' The cost of treating those infections is at least $21 billion annually.

Jim O'Neill, who headed a UK government review of the drugs industry believes that Big Pharma is in for a watershed moment.

---

* In the USA there's even a website called Dollars for Docs where you can track industry payments to individual physicians.

'Somebody is going to come gunning for these guys, just how people came gunning for finance.' (He should know: he used to be the chief economist at Goldman Sachs.) O'Neill warns that, by 2050, drug-resistant infections will kill more people each year than currently die from cancer unless we find some new antibiotics. According to PriceWaterhouseCoopers, without a change in strategy 'no country will be able to meet the healthcare needs of its inhabitants by 2020'. Dr Margaret Chan, Director General of the World Health Organization, puts it bluntly:

> 'A post-antibiotic era means, in effect, an end to modern medicine as we know it.'

We need more drugs, and we need them fast.

'You have to ask yourself *why* we have nearly 5,000 failures to every one drug that we get?' says Samir leaning forward. 'It's no good saying, "because drugs are hard" No! The *process* is wrong!'

I've come to Delhi to see his alternative.

If you want to find out which agents and publicists look after various celebrities and artists, a good place to start is the website Who Represents? Rendered as a web address this becomes whorepresents.com, which without knowing the context might suggest a somewhat different service. The problem of differing interpretations arising from the same sequence of letters is also found in the world of genetics – but here the consequences are far more serious.

When we hear about the genetic code of an organism being 'sequenced', it means just that. The order of the molecules in the DNA is catalogued and written down using the letters A, T, C and G (representing Adenine, Thymine, Cytosine and Guanine

– the four compounds that nature uses to record all her recipes). If you saw your own genetic code represented this way, you'd be looking at roughly three billion unspaced As, Ts, Cs and Gs. Our ability to *sequence* genetic code doesn't translate into us *understanding* it. Anshu gives a very author-friendly analogy: 'A genome sequence is like the book you're writing, but with every full stop and comma removed, with every paragraph running into each other, with no chapter information, index or table of contents. It's just a string of characters, end to end.'

To make sense of that string of characters and slowly unpack the words, sentences, paragraphs and chapters would take a particular set of talents, a great deal of time and a huge dose of patience. A similar task, but an order of magnitude more complex, faces researchers presented with a freshly sequenced genome. Making sense of that sequence – working out where particular genes begin and end, which bits of a cell they are in-structions for, the role of each particular bit, and how all those bits interact – is called 'genome annotation'. Itisa*much*harder-jobthanmakinganormalsentenceoutofasequenceofsquashedto-gethercharacters. Now imagine doing it with the four million consecutive letters that make up a tuberculosis genome. Which is in a foreign language. That nobody really speaks.

Genome annotation is not for the faint-hearted. It requires teasing out the relevant information from myriads of scientific research papers and tying it all together into something approaching a coherent picture. Lincoln Stein, head of Informatics and Bio-computing at the Ontario Institute for Cancer Research, compares it to the interpretation of ancient religious texts:

> 'For thousands of years, rabbis have laboured over the text of the Torah, seeking to make this cryptic, uneven and internally

contradictory text into a coherent system of law, and storing
this commentary into an annotated version of the text, known
as the Talmud. Over time, the amount of annotation in the
Talmud has greatly exceeded the original text – each line of
the Torah is now surrounded by layers of commentary in an
onionskin fashion. So it is with the genome.'

Samir had a plan for his assault on tuberculosis, but it hinged on developing a comprehensive (in fact, the world's *most* comprehensive) annotation of the bug's genome. The good news: there is an awful lot of information out there – by Samir's estimate around 45,000 scientific papers on tuberculosis and its variants, (over half published since 2000 'when Bill Gates started funding'). The bad news: there is an awful lot of information out there. Samir estimated it would take *three hundred* years' worth of work for one person to coalesce it all into a useful result.

Given the rapid advance of drug-resistant tuberculosis, that wasn't good enough. But how could he hope to achieve this Herculean task in any reasonable time frame? Initially he tried it in-house. Along with colleagues Srinivasan Ramachandran and Birenda Mallick, he developed some software to help speed the process, but progress, even with the institute's dedicated and knowledgeable staff, was painfully slow.

So, instead, he decided to get a bunch of amateurs to do it.

'**We knew there were many students around India who could** probably do this work,' Samir says. 'So we thought why not get them all on it? Why not *crowdsource* the annotation?'

Samir tasked Anshu and colleague Vinod Scaria with building a platform to enable just that. The result was Connect2Decode (C2D), a kind of Wikipedia for the tuberculosis genome – an online community space where you could be assigned a task (e.g. a paper to read) and then submit your suggested annotations. The platform was open to students and teachers alike, with Anshu initially believing it would be easier to encourage and co-ordinate student efforts 'if their teachers were mentoring them'. She approached faculty heads to get them excited by the idea but was met with 'crazy levels of bureaucracy' and a barrage of scepticism. 'We were told,' recalls Anshu, 'students don't know how to annotate, you will get a lot of crap and there's no way to do quality checks.'

So Anshu and Vinod started contacting students directly, handing out their own mobile phone numbers and saying, 'We are available 24/7 for questions.' That's when the project really kicked off. 'I can tell you students took our promise to be available any time of the day or night seriously! We were getting phone calls at two in the morning, five in the morning, eleven at night!' That's because those eager to contribute were working on the annotation during free time, outside of their normal studies.

Amazingly, contributors averaged six hours a day on the project for no financial reward. In many cases, groups of annotators got together physically, exchanging tips and techniques. Local team leaders arose entirely on merit. 'In two groups the team leader was an undergraduate while team members were lecturers.' Over a few months, the project identified 114 hard-core 'super-annotators', including one particularly brilliant individual who will enter our story shortly.

Crowdsourcing doesn't just happen. There's a lazy perception in our increasingly online and connected age concerning 'the wisdom of crowds' (a term popularised by James Surowiecki's book of the same name) – that you just throw a problem out to the masses and sit back while an elegant co-evolved solution emerges effortlessly in front of you. To be successful, however, crowd-based projects need intelligent and ceaseless oversight. Anshu gives me a quick masterclass in how to do it right.

First of all, you need to locate the *right* crowd. Then you have to get in touch with them – which, as the team found out, might mean bypassing some gatekeepers. The task you're putting before the crowd must be unambiguous – everyone has to know exactly what the task is and why it needs doing. The process – how contributors' efforts will be submitted and integrated into the whole – needs to be clear. ('That's why we had lots of tutorials on the site,' says Anshu.) Deadlines are important, too, to keep people focused. 'We had strict time limits that if you didn't meet, the system locked you out,' Anshu tells me, meaning the initial 1,200 contributors were quickly whittled down to 350. Anshu is also clear that feedback mechanisms must be easy (both within the crowd, and between the crowd and the project owners) and, if you're in charge, you must respond quickly and intelligently to questions and challenges. It's why they gave out their phone numbers, and it's also the requirement that took the biggest toll. Anshu tells me she 'hardly slept' during the annotation project.

The result, however, was extraordinary – the most comprehensively annotated tuberculosis genome in history, completed in *just four months*, quickly published online, free, for anyone to use.

That ruffled a few feathers.

John Quackenbush, Professor of Computational Biology and Bioinformatics at the Dana–Farber Cancer Institute in Boston, was one of the first nay-sayers. 'It is unfortunate', he pronounced, 'the Indian group made these claims before outsiders had a chance to review the students' data. The worst thing you can do to your country is to oversell your science.' Actually, thanks to the diligent design of the platform, the data *had* been reviewed. Each annotation was scrutinised five times with participants regularly being asked to check and re-check each other's work (and being given kudos for the most corrections). This, however, didn't satisfy the scientific establishment's accepted benchmark of proof: your findings printed within the pages of a 'peer-reviewed journal'.*

Quackenbush's comment featured in *Nature* alongside biologist Pushpa Bhargava's assertion that the idea 'that students reliably annotated the genome so quickly' was 'simply hilarious'. In response, Rajesh Gohkale, a colleague of Samir, commented, 'All the work is available in an open portal; what is the point of publication? *Everybody* can review it.'

That didn't go down well either.

During my trip to Boston I'd learnt, thanks to the work of meta-researcher Dr John Ioannidis, that a lot of what passes for published medical research is riven with bad practice, bias and bravado. Anshu and Vinod's crowdsourced and crowd-*validated* method was arguably *better* science than most of what makes it into scientific journals. In an echo of PatientsLikeMe's communities outperforming physicians, dedicated crowdsourced students were delivering far more robust results than the 'experts'.

When Samir, Anshu and the team did get around to publishing their results as a 'proper' paper, they included something new, that both validated their approach and got a lot of attention. With the most comprehensive tuberculosis genome annotation in the world now to hand, they were able to identify *seventeen* potential weaknesses, or 'targets', in the bacterium. (In the world of drug development a 'target' is any part of the organism's biology that, if you interfere with its operation, can do it serious harm.) Five

---

* Publication is only achieved after a team of recognised experts in your field have had a chance to dissect your methods and, if they have the time and inclination, replicate your results, whereupon (if you've passed muster) your paper is cleared for publication and added to the recognised canon of scientific literature.

of those targets had already been validated elsewhere, making the team confident their unique crowdsourced and super-fast method of genome annotation was accurate. Announcing *twelve* previously unknown and potentially promising holes in tuberculosis' armour was a big deal.

The paper had fifty-four named authors ('anyone who contributed more than 1% of the annotations got a name check,' explains Samir),* the names listed in order of most annotations to the least. Nearly all the contributors were students from Indian institutions, but news of the project had spread, so there were also participants from Germany and Malaysia. Beyond the satisfaction of working on a real-life project and improving their skills, the co-authors received a more immediately braggable reward on publication – having all now co-authored a paper with Professor Samir Brahmachari, India's most senior scientist, something none of them would have thought possible while still an undergraduate. (Another key component of successful crowd-sourcing projects, Anshu tells me, are clear rewards for contributing well.)

At the top of the list was student Rohit Vashisht, who'd participated from the Indian Institute of Science in Bangalore. It was clear to the team Rohit was something special and so he got an extra reward – an offer to become one of them and relocate to Delhi. In fact, he's been sitting behind Anshu, Samir and me the whole time they've been talking me through the annotation project, almost invisible. That's not because he's small or shy, just incredibly still – as if he's constantly thinking, which, it turns out, isn't that far from the truth.

Rohit later tells me of his two key inspirations. One is the 'father of modern computing' Alan Turing, whose

---

* Contributors below this level were acknowledged as part of the 'OSDD consortium'.

extraordinary legacy includes helping to formalise the concept of the 'algorithm', kicking off the field of artificial intelligence and essentially co-inventing and building (with Tommy Flowers) the first modern programmable computer in order to crack the Nazis' Lorenz and Enigma encryption systems. The other inspiration I recognise from the picture above his desk: Srinivasa Ramanujan, one of the greatest and most talented mathematicians who ever lived. To list the achievements of this self-taught genius would take an entire book,* but if I compared his influence on mathematics to that of The Beatles' influence on pop music, it wouldn't be outlandish.

Rohit was just the guy Samir needed next.

A great deal of scientific endeavour involves isolating a particular phenomenon and understanding it in its entirety, a task generally made easier by selecting smaller and smaller targets of study. (By example, the University of British Columbia in Vancouver's study into how herrings fart to communicate, I kid you not.) Samir, by contrast, is an advocate of 'systems science' – putting all that isolated knowledge back together in some kind of coherent framework, trying to divine the 'big picture'. The trouble with systems science, of course, is that it can get horrendously complex incredibly quickly, especially when it comes to biology. The number of interactions in even the simplest bacterium (e.g. tuberculosis) is staggering.

To get an idea of just how insanely complex I mean, try typing 'metabolic pathway map' into Google. What you'll get are schematics that attempt (in the larger examples) to show

---

* ...and that book would be Robert Kanigel's *The Man Who Knew Infinity*.

biological systems in their entirety. They are immediately overwhelming – a clutter of tightly packed rectangles and circles connected with spider's webs of lines and arrows that could easily take up an entire wall if printed at a readable size. Even if you know what you're looking at, they're hard things to get along with, or make sense of. The diagrams ('the way we've represented this knowledge for the last twenty years') are unwieldy and, Samir argues, obfuscating. They may attempt completeness, but the cost is clarity. By analogy, imagine trying to draw a diagram of all your interactions with everyone you've known for your entire life. Imagine you somehow achieve this, creating a hugely complex and intertwined schematic. Now imagine you're off to see a psychotherapist (almost certainly to discuss a childhood dominated by an unhealthy obsession with diagrams) and in your first consultation you hand them your chart and refuse to talk for the rest of the session. Would you trust the therapist to do a good job? To find meaning in the madness?

Anshu and Vinod's annotation project had thrown up 1,152 chemical reactions of interest involving 961 substances, all regulated by 890 genes or, to put it another way, a complexity-fest. Just *listing* them would take up a good deal of paper, let alone trying to diagram how they all interact in the traditional way. The team *did* create a traditional metabolic map, which, when I looked at it, made my brain immediately baulk.

Samir wanted something better – as he describes it, 'something I can manipulate and ask questions of' – and put out a call for computing undergraduates, asking those who met his criteria (no doubt by arguing with him) to 'convert this complex diagram into a computational model'. He wanted to create a 'virtual' tuberculosis bacterium he could experiment on inside a computer.

'Drug development is stuck in the past,' argues Samir. 'Today we build and test our aeroplanes and our cars in the computer before we build them for real. But we build drugs in the petri dish – trying out different compounds to see if one works. We need to build computer simulators like they have in aerospace; it's much more efficient.'

'You wanted to build a simulator so you could see how the bug "flies"?' I ask.

He laughs. 'I wanted to build a simulator to see how you could *crash* the bug!'

The idea of simulating a bacterium in a computer (an '*in silico* model' in the parlance of computational biologists) is not new, but the insanely interwoven nature of whole biological systems means that creating a simulation for even relatively 'simple' organisms like bacteria is no small task. The bedrock upon which any simulation rests is the accuracy and completeness of your genome annotation. Luckily Samir had the world's best annotated tuberculosis genome on his hands, thanks to Anshu, Vishnu and their crowd of amateurs. But this, of course, brought its own woes. The more detailed the data, the more you have to include in your simulation. How did his new recruits do?

'All I can tell you is they had a miserable time! They couldn't figure out anything!' chuckles Samir. 'Part of the problem was these were computing kids – they didn't understand biology. I needed a way to reframe the problem for them.'

His solution was to ask them to model the movement of packed lunches.

In the world of logistics there's one organisation that has attained almost mythical status. It makes a mistake just once in every sixteen million deliveries and counts Richard Branson and several Harvard Business School professors among its fans. Corporations sign up for the seminars and workshops it offers hoping to learn how, for over a century, it has achieved consistently astonishing performance. So who is it? Federal Express? UPS? No. It's the 5,000 *dabbawalas* of Mumbai who deliver 200,000 tiffins containing home-cooked lunches every day to workers in the city. Even the raging monsoon fails to dent their performance. Not bad for a system which is self-managing (every *dabbawala* is their own boss), engages a semi-literate workforce and relies entirely on bicycles and the public train network. ('We are humble people,' say the *dabbawalas* on their one obvious nod to the modern world: their website. 'We don't know management theories. All we have is decades of learning.')

In a flash of inspiration, Samir saw that there was something in the way the *dabbawalas* had solved the complex problem of delivering 200,000 tiffins a day within the organism called Mumbai that could be repurposed to help his team model the complex inner life of a bacterium. Each tiffin is painted with a collection of coloured symbols: a code for the collection point, a code for the train station where the tiffin will be cycled to, a code for the destination train station and a code for the final drop-off point (including the floor of the building that is the ultimate destination). The *dabbawala* system is a relay, with each tiffin being handed from person to person, each sending it on its way based on what the coding system tells them.*

Bacteria are also complex systems passing things about; proteins, lipids, carbohydrates – a smörgåsbord of cell-stuff that has to be moved from one place to another or, more often, be converted from one form to another. Biologists call these chains

---

* The *dabbawala* system been compared to the 'packet-switching' system that underlies the Internet – where each chunk of data (a packet) carries its own origin and destination data with it – and is sent onwards by 'routers' reading that data and directing it accordingly.

of chemical reactions 'pathways' – production lines mediated by enzymes (essentially biological catalysts, or 'reaction engines' that chivvy up the conversion of one thing into another) that make sure the bacterium gets the end products it needs.*

'So imagine', says Samir leaning forward, 'that each journey a tiffin makes from home to office is like a biological pathway, a chain of reactions in the bacteria, and that each stop on its journey is the place where an enzyme converts one thing to another, and every movement of a tiffin by a *dabbawala* is like the result of the last reaction being carried to the next'. Using this metaphor could, Samir hoped, unlock the way to create a useful simulation of the bacterium. Of course, you'd need a computing genius who just happened to know a lot about the biology of tuberculosis on your staff …

I'm looking at what, on first glance, could be mistaken for a train map that shows the links between three incredibly thin

---

* A common pathway you may be familiar with is the one by which humans process alcohol. As you're downing that glass of beer your liver is busy making a suite of molecular machines (those 'enzymes') specialised for the task of breaking down booze. The enzyme 'alcohol dehydrogenase' first turns the alcohol into acetic acid. After that, the inventively named 'acetaldehyde dehydrogenase 2' turns that acetic acid into acetyl-CoA, which enters 'the citric acid cycle' (another chain of enzymic reactions), breaking it down into water and CO_2, which your body can dispose of. Variations in this pathway between different groups of people mean that some of us can metabolise alcohol better than others. (For instance up to half of East Asians carry an inactive acetaldehyde dehydrogenase 2 gene, hampering their ability to process wine, beer and spirits.) If you don't accomplish these reactions fast enough, too much neat booze finds other places to hang out in your system, including your brain, interfering with cellular processes up there – which is why you get drunk (and, if you drink too much, can't remember doing so). It's also the reason that drinking while pregnant is so dangerous for your baby. Their tiny livers aren't developed enough to make the number of alcohol-handling enzymes required, leaving the booze to wreak havoc with the rest of your infant's developing metabolism.

cities. In actuality it's probably the most complete simulation of a bacterium anywhere in the world.

City A on the left has 961 stations all in a straight line, north to south. Each station, Samir tells me, represents a 'metabolite' – any substance created by the bacterium as it goes about its business. City B in the middle of the diagram has 890 stations, again all in a straight vertical line. These stations represent a bacterium's genes, the recipes for cell stuff. City C, on the right has 1,152 stations, again arranged north to south, each representing a reaction that can take place inside the bacterium. These stations are grouped into common 'pathways' (regularly used chains of reactions).

Thin train lines link the stations in each city with their counterparts in the neighbouring one, the lines appearing or disappearing depending on what state the bacterium is in. You can trace these lines if you want, finding out which metabolites are being used by which reactions at the behest of which genes. (Samir and Rohit call their simulation the 'Systems Biology Spindle Map' – each line of stations being a 'spindle'.) It's still an overwhelming amount of data to look at but crucially, says Samir, 'a computer scientist can ask questions when looking at this picture. Like why are *these* metabolites' – he points to a place on the diagram where many train lines converge – '… connected to so many reactions?'

Straight away you can see which processes it might be beneficial to interfere with. If a particular metabolite is used in lots of reactions, you could use its gregarious nature to damage the bug. Stop that metabolite being made, or interfere with it somehow, and you're going to mightily upset the bacteria (in the same way that closing down a particularly busy station at rush-hour will really annoy commuters and hamper the smooth running of a city). Similarly, if an enzyme is used in several 'pathways', messing with it can turn a good day into a bad one for your bacterium.

Samir and Rohit show me their simulation of a tuberculosis bacterium having an undisrupted day, drawing my attention to 'stations' on the map where there's lots of traffic.

'It makes sense to attack these,' says Samir. 'That would cause the most disruption.'

In fact, this is exactly what a lot of existing drugs do.

'Let's see what happens in the presence of Isoniazid,' says Samir, a reference to one of the current 'frontline' tuberculosis antibiotics (even though it's now over sixty years old).

The schematic shifts to simulate the bacteria's response and it's striking. Even to my untrained eye I can instantly see the bug isn't happy. Train lines are shut down. Reactions aren't happening. Cell-stuff isn't being made. Stations on the network are suddenly isolated. 'We see lots of genes shut off, lots of reactions stop, the growth of the bug goes down. Good!' says Samir. 'But look!' He leans forward and points. 'We find *sixty* other genes are now newly called into action. It's *compensating* as best it can.'

Sure enough, another, smaller, set of train lines have popped up that weren't there before, representing reactions and pathways that were dormant in the happy bug, but have arisen when it's under attack – in much the same way commuters find alternative routes during transport outages. Of the 60 called-into-action genes the model simulates when subjected to virtual Isoniazid, 48 of them are already known thanks to previous studies on the drug. The other twelve, Samir suspects, are 'in such low concentrations we can't detect them in experiments' meaning that the simulation could well be revealing truths that the lab cannot. That's surprising. Most simulations used in medicine are considered to be approximations of lab work. Rohit's simulator, based on Anshu and Vinod's genome annotation, is so good it may be outshining what lab science can currently reveal.

'So, you're showing me how the bug resists the drug?' I ask.

'Yes, we're mapping how tuberculosis becomes drug-resistant,' explains Rohit. 'The simulation can mimic the evolutionary process.' He pauses. 'Other models cannot do that.'

Everyone here is clear. Rohit's ability to take the genome annotations, along with the *dabbawala* metaphor, and then guide a team of 'computer geeks' (as Samir affectionately calls them) to code the mathematics that power the simulation, is a stunning achievement. In the same way Alan Turing and Tommy Flowers used the power of computing to crack Nazi codes so the Allies knew the Germans' intentions, Rohit's simulation can read the mind and predict the actions of the enemy.

For Samir, it's a key weapon in his war against drug development inefficiencies, seeking out more promising potential medicines from the get-go. Why pursue avenues of attack the simulator tells you in advance the bug can compensate for?

'Informatics is cheaper than the experimental world,' he says. 'We don't have $2.6 billion! When there is a shortage of money you have to innovate.'

Having built the simulator, the next obvious thing to do was to give it away, delivering into the hands of anyone who wants it a virtual bug that can be interrogated without risk of harm to the investigators. It opens a new area of tuberculosis research to people whose labs don't have the necessary bio-safety measures in place to deal with the real thing.

'We give away everything,' Samir says. 'That's why we call it *Open Source* Drug Discovery' (or OSDD for short).

In the world of Open Source software, there's an oft-quoted mantra – 'given enough eyeballs, all bugs are shallow' – meaning that if you get plenty of programmers to look at some misbehaving code, the error (the 'bug') will reveal itself much sooner – and the resulting fix will probably be better than from a lone geek. OSDD is proving that it's true not only for software bugs, but real ones too. It's Samir's challenge to the existing drug discovery model. Fast, innovative, cheap, collaborative across borders, without ego or any protection of 'intellectual property'. It's everything the industry that takes $2.6 billion to make drugs for the rich isn't.

'We don't believe in the Western concept of knowledge being proprietary. I'm not in this for wealth but because I would like to see a reduction in death,' says Samir. 'If you look at centuries of innovation, from architecture to sex, things were given away. The Kama Sutra is Open Source!'

In short: more sex, less death. I like the guy.

Fixing the problem of drug-resistant tuberculosis would be tough enough if it was just one strain of bacteria, but it isn't. Tuberculosis is more like a large mafia family, a bunch of evil bugs with different characteristics, some more drug-resistant than others. Finding a way to kill one of them won't necessarily affect the others.

Armed with their new tools, Samir, Rohit and their colleague Dr Divneet Kaur (whom I'll meet shortly) began asking themselves if they could find a line of attack that might affect *every* strain. Luckily for them, researchers at the Theodosius Dobzhansky Centre for Genome Bioinformatics in St. Petersburg, Russia, had been busy compiling the 'Genome-wide Mycobacterium Tuberculosis Variation database' – a catalogue of 1,849 different tuberculosis genomes.

Immediately the team descended on the data. Were there any genes that showed no variations across all of the catalogued strains? The theory was that if they could find such a gene it must be essential for the operation of the bacteria in exactly that unmodified form. Any change in the gene, even a single letter alteration to its DNA, was likely to spell disaster for the bug, given no strain with that change had survived to make it into the Russians' database. Finding such a gene could give the researchers the tantalising possibility of creating a catch-22 for *all* tuberculosis bacteria, even the drug-resistant strains. If you could create a drug that interferes with the work of any of these 'invariant' genes it had a good chance of killing the bug. And if the bacterium tried to evolve around the drug it'd die anyway.

'Did you find any?' I ask.

'We found eleven that were identical across nearly all tuberculosis strains,' Samir tells me.

Whoah. That's *eleven* potential Achilles heels, clues for medicines that won't just harm one particular strain, but could potentially spell a tuberculosis genocide. And, sure enough, when Rohit knocks out any of those genes in his simulation the results are promising.* It's all bad as far as tuberculosis is concerned. But the team are hardly celebrating. Identifying key targets like this is crucial, but getting from that point to drugs that actually work is far from easy. Anshu tells me a medicine needs to be 'druglike' – an industry term meaning it not only has to be an effective killer, it also has to be an incredible navigator, impersonator and trained assassin.

Even if you have a drug that can kill the bacterium there's no guarantee your body can get it to where it needs to be. You

---

* Anshu and Rohit also cross-referenced those genes with the human genome to make sure the targets were safe for humans. Given that every living thing has a common ancestor, there is a lot of commonality between the genes of different species including bacteria and humans, so this cross-referencing is an important step in making sure you don't create a drug that takes out the patient before it kills the bacterium.

might metabolise it *before* it reaches the place of infection for instance. This is why many drugs come in capsules, giving the stuff inside the protection it needs to reach a point in your gut where it can be absorbed into your system in the correct amount. (It's also why some tuberculosis drugs are delivered by an asthma pump. Because tuberculosis is a disease of the lungs, it makes sense to bypass the assault course of the intestines.) Even if your body *can* get the drug to the bug it doesn't mean the bacterium is going to welcome it with open arms. Your drug needs a *disguise*. You've got to create something that the bug will let inside, a kind of molecular Trojan Horse. On top of that, once it gets into the bacterium the drug has to interfere with a very specific part of the bug's machinery, say a particular enzyme, binding with it in a way that successfully interferes with its target. That is no easy task, either. You've got to find the right bit to latch onto (hard), actually latch on to it (harder) *and* latch on in a way that does some useful damage (harder still – it's not like the enemy puts up a sign saying, 'Punch me *here* and in *this way* for best results').

'So a successful drug is like James Bond trying to shoot the lone nut with a bomb running around in a crowd of thousands of innocents from, like, a mile away?' I ask.

'Yes,' says Anshu. 'You are absolutely right. That's the challenge.'

It's why successful drugs are often called 'the magic bullet' by those in the industry.

'It's time to meet Divneet!' says Samir.

We move to the Institute's bustling cafeteria. There's a definite air of bonhomie about the place. Over incredibly sweet tea,

Samir introduces me to Dr Divneet Kaur who, according to the OSDD website, designs 'non-toxic ligands for targets identified by systems-level analysis of metabolism in Mycobacterium tuberculosis'.

Samir summarises: 'She makes magic bullets.'

Divneet tells me her job is 'a chemist's dream' and she's clear that the OSDD project is doing something radical when it comes to drug development.

'I did my PhD in organic chemistry, which includes no systems biology at all.'

'No thinking about the organism in its entirety?'

'No, we just make molecules.'

She's says this as if 'making molecules' is like baking biscuits, but it's no easy task and different chemists often specialise in particular types of drug, becoming nano-scale craftspeople, working out how to chain atoms and smaller molecules together to create very specific compounds. I wonder aloud if each of these chemists is a bit like a winemaker who knows their own soils, grapes and climate – and juggles them to make their own particular wine?

'Yes, yes,' agrees Samir. ' Organic chemists are like winemakers in this context. And they can also make cocaine or Viagra... depending on their specialism,' he adds, chuckling, a reference to India's rich diversity of organic chemists, not all of them behaving strictly legally. 'Many years back, Pfizer's vice president called me. "There are 174 Indian companies making Viagra!" he complained. Why is he upset? Because the Indian clones were costing him $200 million of revenue!'

You can tell Samir finds it hard to be sympathetic to Pfizer's pain. India's 'generic' drug industry is the best in the world, making cheap versions of branded drugs (often still in patent, much to the chagrin of the big pharma companies), and it'll prove itself very useful when it comes to manufacturing the tuberculosis medicines he wants to create.

Right now, Divneet is particularly excited by one of the 'invariant' genes uncovered by the team's analysis of the Russian database, 'because it's involved in folate biosynthesis.' She says this in the way I might talk about a rediscovered Pink Floyd track.

'And that's exciting because ...?'

'Humans don't make folate. Bacteria do; in fact it's *essential* for them.' (This is probably one of the reasons the gene involved can't vary itself without the bacteria going extinct.) 'That makes it a very promising line of enquiry,' she says. 'There's a high chance of it being very effective against tuberculosis, without side effects for the patient.'

Tuberculosis medicines have something of a reputation for unpleasant side effects. Take the drug Rifampin, which, although it attacks the bacterium, is also a 'well-known cause of clinically apparent, acute liver disease that can be severe and even fatal', thanks to its toxicity.

In the language of chemistry, the molecules Divneet designs are called 'ligands' – made to look attractive to bacteria which then, once inside, attach themselves to crucial proteins and, in doing so, disrupt the bacteria's metabolism. Like everything in drug development, Divneet's job is not easy. Ligands might fail in all sorts of ways: not getting into the bug in the first place, not attaching to their target, attaching to the wrong part, or attaching in a way that doesn't inhibit it enough. But at least she knows what she's aiming at.

She points to Rohit, ever silent, sipping his tea with us.

'Rohit told me what I need to go after. Now I'm trying to come up with the right molecule, something humans can take...'

'... but bacteria hate?'

'Yes.'

It's a marked contrast to the traditional procedure for finding ligands, which is simply to take all the molecules you think might work and throw them, one by one, into separate dishes that contain your bug. 'It might kill the bug,' says Divneet, 'but you've no clue how, and it's probably going to turn out to be toxic to the patient.'

Earlier in the day Anshu had explained how the compartmentalisation of the drug development process means few people in it have an appreciation of 'the big picture', and often don't talk to each other. Divneet picks up the point.

'For the chemists, they're only looking at the molecule, they often have no clue as to its eventual target, whether that target is also present in patients … All these things are ignored and are picked up at a later stage.'

It's an incredibly inefficient way to go about things and one contributor to the drug industry's 5,000:1 (0.02%) success rate, but without the kind of precision targets Samir and Rohit's model can provide, it's become standard practice. In the back of my head I can hear the creators of Eroom's law making their point that pharmaceutical corporations, instead of innovating, have developed a 'throw money at it' tendency – increasing budgets to do the same stuff less efficiently.

'It's expensive,' says Divneet, 'and it doesn't work very well.'

We leave the cafeteria for a tour of the rest of the Institute, during which I meet Samir's extended team, all fully behind the idea of a drug development system that's more targeted, uses the crowd where appropriate and is completely Open Source. I find it heartening that over half (by my estimate) are women, at all levels of the organisation, from administrative positions to senior research posts. This lack of sexism is in marked contrast to Indian society generally.* By the time I leave the Institute at eight o'clock that evening, I am exhausted. I see no such tiredness in the researchers. All the lights are on and the labs and offices continue to bustle with activity. Samir will be here well into the night.

Samir had the idea for OSDD in 2007, just two years after the term 'crowdsourcing' was coined in *Wired* magazine. His original intent was for:

> 'A decentralised web-based community-wide effort, where students, scientists and technocrats, universities, institutes and corporations [can] work together for a common cause [to] bring down the cost of drug discovery significantly by knowledge-sharing and constructive collaboration.'

At the time he was ridiculed for thinking such an approach could yield useful results. Less than a decade later the behemoths of the pharma industry, shackled by their own inefficiencies are (at least partially) admitting defeat. Andrew Witty, CEO of GlaxoSmithKline, acknowledges the blockbuster model is flawed, based as it is on 'finding a needle in a haystack right when you need it' – hardly a sustainable business strategy, especially when the 'blockbuster drugs' of yesteryear are falling out of patent protection and the industry is finding the replacements to prop up the balance sheet harder to come by, and costlier to develop. He also concedes that his industry's oft-quoted

---

* Out of 142 countries ranked for gender equality in the World Economic Forum's *Global Gender Gap Report 2014*, the country comes 114th overall, scoring particularly badly on 'economic participation and opportunity' (134th in the list, just below Lebanon). India's female labour participation rate was 27% in 2013.

astronomical R&D figures are little more than a figleaf used to cover up serious problems with efficiency. 'If you stop failing so often, you massively reduce the cost of drug development,' he told a healthcare conference in London.

In 2010, GlaxoSmithKline, the Genomics Institute of the Novartis Research Foundation and St Jude Children's Research Hospital in Memphis released into the public domain the details of over 20,000 compounds that they knew were in some way active against malaria. By December 2011, Medicines for Malaria Venture (MMV), a Swiss not-for-profit, had selected four hundred of the most promising and now makes them available to researchers, on request and free of charge, in 'the Malaria Box'. What they ask for in return is that any resulting findings 'be published and placed in the public domain to help continue the virtuous cycle of research'. At the same time, they kicked off funding for the Open Source Malaria project based out of the University of Sydney, which is currently working on a promising set of compounds Pfizer has released to them.[*]

Samir is partnering with Open Source Malaria and publishing all findings online. Simultaneously the TB Alliance has asked Samir's team to oversee human trials for a new tuberculosis drug combination they've been working on. Anshu continues to build platforms and communities for open collaboration as OSDD expands its remit to target not only tuberculosis and malaria but also leishmaniasis, a 'neglected tropical disease' with 1.3 million new cases annually. Today, OSDD claims participation

---

[*] It's worth noting that MMV has helped introduce four new malaria medicines to the market since its formation in 1999, which, by pharma standards, is swift work – and they continue to push forward. As I write this, news of a promising antimalarial compound developed in conjunction with Dundee University has popped into my inbox.

from nearly 8,000 individuals in 130 countries, across 75 organisations – all attempting to do something about those ungodly drug development costs. Not bad for an idea pretty much everyone said wouldn't work.

Of course, most of the lines of enquiry Samir's team are following will turn out to be dead ends. Indeed, a few months after my visit he sends me an email listing a good number of stalled investigations. Drug development from whatever angle you look at it is incredibly tough, with so many hurdles at which you could fall. But that's not the point. The point of OSDD is to reduce the ratio of failure to success, to get the figure 'from 5000:1 to 100:1'.

A direct message on Twitter. It's from Rohit. Since I left Delhi we've traded the odd update on a new film about the mathematician Ramanujan, starring Dev Patel, that we're both looking forward to. This message, however, is about something else. It reads,

> *'I figured out how a cheap drug that is used to treat diabetes can be used for TB.'*

That stops me in my tracks.

Continual refinement of the *in silico* model (as more experiments feed information back into the OSDD process and annotation continues to be done) has unveiled one of tuberculosis' drug-resistant tricks, an enzyme that gets called into action when existing medicines try to shut down the bacteria's energy generation system – a new train line on his 'spindle map' (or, as Rohit pithily puts it, 'directional re-routing of metabolic fluxes through NAD de novo biosynthesis pathway and respiratory chain complex'). It turns out that there is *already* a drug that could target the enzyme in question, approved for use by diabetes sufferers. In summary, the OSDD process could have found a new tuberculosis therapy for a tiny price tag in comparison to pharmaceutical industry norms, the equivalent of finding a priceless heirloom in the attic. In fact, to date the *whole* OSDD project has spent less than $15 million – peanuts in the world of drug development (it's just under 0.6% of the industry's $2.6 billion per drug).

Interestingly, at nearly the same time as Rohit was submitting his findings to the *Journal of Translational Medicine*, a paper in a competing journal came out with pretty much the same findings. There was, however, one big difference. While Rohit came to his conclusions sitting largely alone with his simulation, the other paper listed sixteen authors across nine institutions who'd conducted in-depth experiments with a variety of drugs on some rather unfortunate mice, work they wrote up in a paper totalling 10,000 words. Rohit's write-up was less than 1,000 words – detailing what his simulator had told him. Same result, but an order-of-magnitude difference in efficiency.

His next message reads, 'Just a small update. I am happy to tell that I got a post-doctoral position in Stanford University', easily one of the most prestigious medical research institutes in the world. Following my congratulations he writes, 'It's a great place … with highest number of Turing Award winners', referring to the 'Nobel prize' of computing, awarded each year by the Association for Computing Machinery. Perhaps they've found another potential winner. After all, Turing used the power of computing to shorten World War II by up to three years, helping to save an estimated twenty-one million lives. Rohit's war is on disease – and his algorithms could potentially help to save billions.

During my time in Delhi, Samir had urged me to look up the writings of the Indian philosopher-monk Swami Vivekananda.

'You'll enjoy it!' he exclaimed with typical enthusiasm.

As I head to my next destination, a trip to the northern agricultural province of Ranchi (where I'll find an impossibly low-tech solution to one of the world's grand challenges), I take his advice and find myself reading a letter by written by Vivekananda to one of his disciples who had sought the swami's guidance on a new project. He wrote:

> 'Go on bravely. Do not expect success in a day or a year.
> Always hold on to the highest. Be steady. Avoid jealousy and
> selfishness. Be obedient and eternally faithful to the cause
> of truth, humanity, and your country, and you will move the
> world.'

4 RICE WARS

'The test of our progress is not whether we add more to the abundance of those who have much; it is whether we provide enough for those who have too little.'

– FRANKLIN D. ROOSEVELT, US PRESIDENT

I'm getting slightly worried that I may be about to die. Our car has just passed onto the wrong side of the dual carriageway and is heading directly into oncoming traffic.

'Don't be alarmed,' says our host Yezdi.

This strikes me as unhelpful. Can he not see the fast-approaching small goods truck directly in front of us? I thought drivers in Delhi were cavalier, but here in rural Jharkhand our man at the wheel seems positively suicidal. But our host is right. The lorry ahead calmly moves out of our way, without even a toot of its horn.

'The turn-off is on the wrong side of the carriageway and there is no other way to get to it except to go the wrong way for a bit, so local people expect to see oncoming cars.'

With road planning like that it is perhaps no surprise that India has developed an unspoken highway code based primarily, as far as I can work out, on chaos theory. As we exit the carriageway (much to my relief) it strikes me that my moment of terror is something of a metaphor for the problem that's brought me to Jharkhand. Humankind is currently playing chicken with an oncoming challenge that, if it hits (and it will unless we do something soon), will destroy a good chunk of what we've come to call civilisation. In short, our global food system is heading for collapse. If it fails billions will starve.

I've come to Jharkhand to learn one way we can avoid disaster.

My visit here is the result of serendipity. I already had soil and crop scientist Dr Erika Styger in my sights but by luck she's in India the exact week after my trip to Delhi as a guest of Krishi Gram Vikas Kendra (Agriculture Rural Development Centre), and she's generously invited me to join her and her friend and colleague Dr Gaoussou Traore as they visit the fields and villages of Jharkhand. Erika's tall, Swiss and intense. Not in a bad way. In fact, she's easy company and laughs readily, giving her broad and thoughtful features a welcome respite from their default expression, which seems to say 'prove it'. She questions everything. If someone makes a claim, she'll instantly want them to back it up with evidence *and* explain the methods by which they acquired that evidence. Gaoussou, by contrast, is the very definition of benign composure. While the rest of us will struggle with the uneven terrain of the fields and villages we'll visit here, his tall frame, clad in the flowing fabric of his *boubou* robe, seems to glide effortlessly. He nods a lot, often while making appreciative 'ah' sounds, and when he chuckles (which he does regularly) it's like somebody is ringing a bell to announce a small, exclusive party to which you've been lucky enough to be invited. Gaoussou coordinates agricultural projects across thirteen countries in West Africa from a base in his native Mali. Whether the demands of such a large job have taught him the benefits of serenity, or an inbuilt composure made him perfect for the job, I can't tell.

As we arrive at our destination Erika tells me that her and Gaoussou have flown in from a conference in Thailand, where at least one delegate told her 'they wanted to put poison in my coffee' and both were kept on the fringes of the event. It's a response she's used to. The work of her institute, based at Cornell University, isn't universally liked, having been dubbed 'nonsense', with 'no empirical or theoretical basis', its findings 'unsubstantiated' and 'miraculous' (not a compliment). In what's becoming a common thread on my journey, it appears those people who question the status quo can expect rough treatment from the incumbents. All that said, she seems in remarkably good spirits.

We've come to KGVK's training campus, the hub of an organisation set up by the Usha Martin manufacturing group to promote 'sustainable and integrated rural development'. The company's founder Brij Kishore ('BK') Jhawar 'felt we had to do something for the community when it set up here in 1961', explains Yezdi, who shows us around, 'but he was an industrialist and, at the time, didn't really understand social dynamics or rural communities. If we're honest, part of the motivation was to make sure there was no social unrest in the area where he was operating.' Today, BK dedicates his efforts full time to KGVK and the 400 villages the organisation works with.

We start in the dairy, a small training operation with eight or so cows, all listening to Indian classical music. It's been scientifically proven that relaxed cows produce more milk and the right music has a calming influence on them. (The researchers who discovered this couldn't resist calling their work a 'Moosic study'.)

'Nothing too fast,' says Yezdi. 'No hip hop!'

Our next stop is the campus farm, used for training and research, where Sudhir Paswan, a young man who will come to amaze us all, guides us to a plot of rice. To my untrained eye there's nothing special about the small crop we're looking at (there again, before this moment I've never seen a rice plant in my life) but for Erika it's a different story.

'Oh wow!' she exclaims. 'Oh *wow*!!'

'Ah, she is verrrry happy!' laughs Gaoussou.

Erika steps off the track into the plot plunging her hands between the stems or 'tillers' of the plant. She looks up at Sudhir.

'One seedling?'

Sudhir is smiling. 'Yes.'

She turns to Gaoussou. 'This is extreme, no?'

Before he can answer, Erika is back on the path, catching up with Sudhir, who, wanting to show us more of the farm, has moved on. I can hear her bombarding him with questions: 'What age are those plants? When did you plant them? What are you fertilising them with?'

Gaoussou is smiling, as he seems to most of the time. 'When Erika starts to talk about rice, she can go on for hours!'

His chuckle mixes with the early evening chorus of birdsong and, suddenly, I feel very happy to be alive.

The following morning the sun is shining as we get in our car ready to drive out for a series of field visits. As we set off, Yezdi tells me a little more about the state and its challenges.

By any analysis, most of the citizens of Jharkhand are dirt poor. The Oxford Poverty and Human Development Initiative ranks the state and its neighbour Bihar as two of the five poorest regions in South Asia (the other three being South and West Afghanistan, and Balochistan in Pakistan). Over half the population (32 million) live below the poverty line. That poverty is particularly unjust when you realise the state is one of the richest in India when it comes to natural resources, but the bountiful reserves of coal, iron ore, copper ore, mica, bauxite, limestone and uranium haven't translated into wealth for the locals. Whilst the industries that exploit these resources might employ local residents, it is citizens elsewhere, distant shareholders and government officials, that have enjoyed most of the spoils.

'Various governments over the years have either misused or misdirected resources,' explains Yezdi, and this is part of the fuel that fires India's 'greatest internal security threat' (according to former Indian premier Manmohan Singh) – the nearly fifty-year-long Naxalite Maoist Insurgency. Either side of my visit the regional newswires light up with tales of ambushes of local police. It's estimated the troubles have claimed over 13,000 lives, mostly civilians. 'There are many areas we can't operate in because of the Naxalites,' Yezdi tells me. 'It's too dangerous.'

Another problem facing Jharkhand is a lack of groundwater, one of the reasons just 6.6% of the total cropped area in the state is 'lift' irrigated (using pumps and tubewells). KGVK is working with the government to create an irrigation infrastructure, but there are challenges.

'To bore a well, you need government approval because groundwater is so scarce,' says Yezdi. 'Even if you have a licence you sometimes have to go 600 metres down. That's expensive and puts it out of reach of most farmers. Besides, we discourage practices that have a negative effect on the water table. We don't want them to flog a dead horse.'

'So what do they do?' I ask.

'They wait for rain,' says Gaoussou, as the morning sun begins to properly bake the fields we're driving past. 'They wait for rain.'

That reliance on the clouds makes what I'm about to see here in Ranchi all the more remarkable.

In 1961, the world's population was just over three billion and the average number of children born per woman was five, presenting humanity with a problem. At the time the top twenty agricultural commodities yielded about 2.1 billion tonnes of food a year. Averaged across the globe, food calories available per person came to about 2,300, about enough for everyone if they were distributed equitably (which, of course, they weren't); but, with the population bomb continuing to explode, the world was staring down the barrel of mass starvation.

Amazingly, by 2011, when we hit seven billion souls, the tonnage of the top twenty commodities had tripled, more than keeping pace, and per capita daily calories available (averaged worldwide) were pushing 3,000 – although still tragically unevenly distributed. In the same period, deaths from famine plummeted 200fold, while the percentage of undernourished people fell, and indeed continues to do so. Today, the UN classify 11% of the world's population as undernourished, down from 18.5% at the beginning of the 1990s.

Agriculture's saviour was the 'Green Revolution': a combination of specially bred 'high yield crops', chemical fertilisers (notably potassium, nitrogen and phosphorus), synthetic herbicides (for controlling weeds), pesticides (for deterring insects) and increasing mechanisation (allowing fewer people to farm more land) – all of which transformed food production and our modern world as a result.

The man most associated with this revolution is Norman Borlaug. A multi-disciplinarian conducting research into plant genetics and breeding, entomology (the study of insects) and agronomy (the science of crop production and soils), he is often given the accolade of being the first person in history to save a billion lives, and is one of only six people in history to receive not only the Nobel Peace Prize, but also the US Presidential Medal of Freedom and the Congressional Gold Medal. Not bad for a boy who grew up attending a one-room prairie school in Iowa and failed his university entrance exam. Instead of becoming a high school science teacher, his initial ambition, Norman ended up with his work being taught in pretty much every high school in the world.

Borlaug has been described as 'the greatest human being you've probably never heard of.' However, he's also been accused of being a key architect of 'rural impoverishment, increased debt [and] social inequality', with political journalist Alexander Cockburn going so far as to write that, 'Aside from Kissinger, probably the biggest killer of all to have got the Peace Prize was Norman Borlaug.' Three days after his death in 2009, the *Guardian* was asking, 'Has there ever been a person in human history whose legacy has pivoted so precariously on the fulcrum between good and bad?' How can the man most readily associated with the revolution that fed the world be such a divisive character?

Borlaug's work (and that of those he inspired) centred around creating high-yielding, disease-resistant strains of our 'global food security crops': wheat, rice and maize (so called because between them they provide roughly half the world's plant-derived food energy). But the seeds of Borlaug's new wheat varieties did not grow higher-yielding plants in and of themselves. To get those you needed to give the plants extra inputs of fertilisers and water to which they are far more responsive than natural varieties. There's no 'waiting for rain' with the Green Revolution; you need irrigation. And that's turned out to be a big problem.

Take the Indian state of Punjab, for instance, which became known as 'the breadbasket of India' thanks to its enthusiastic take-up of Borlaug's seeds and associated farming methods. Production of wheat, rice and other crops in the state rocketed from the 1970s, but so did the number of tube wells needed to satisfy the soaring demand for water. The result was that the water table (the level below which the ground is water-saturated) began to plummet. Today the Indus Basin aquifer, on which the Punjab relies, is the second most stressed in the world and the bigger farms are only surviving by using adapted oil-drills to reach water 300 metres underground. Many smaller operations have gone out of business or become saddled with debt thanks to the expense of the machinery – a major contributor to farmer suicides, of which there were 5,650 nationwide in 2014. It's a vicious circle. The crops' voracious thirst insists on ever deeper wells which drain the water further out of reach. Tushaar Shah of the International Water Management Institute in Gujurat puts it starkly: 'When the balloon bursts, untold anarchy will be the lot of rural India.'

China is another example. The nation embraced the Green Revolution eagerly following the great famine of 1959–61 (which is estimated to have killed 45 million people). Today many farmers are drilling as deep as their desperate counterparts in Punjab. The water table around Beijing has dropped nearly 13 metres since 1998.* Meanwhile California has introduced mandatory water rationing for the first time in its history. A well deep enough to reach the ever-receding H_2O

---

* The government's response is one of the largest and longest engineering projects in history: the South-to-North Water Diversion Project, which will link together the mighty southern rivers (Yangtze, Yellow River, Huaihe and Haihe) and divert 44.8 billion cubic metres of water a year to the thirsty population centres in the north. Completion is estimated in 2050, at a cost of $62 billion.

can cost up to $45,000 for a residential system and $750,000 for an agricultural one, and the waiting time for a new well is two years.

Lester Brown, founder of the Earth Policy Institute, warns that 'countries containing half the world's people are over-pumping their aquifers', meaning the world is facing 'a water-based food bubble'. As Alexandra Richey, from NASA's GRACE project, which uses satellite data to monitor our groundwater, asks, 'What happens when a highly stressed aquifer is located in a region with socio-economic or political tensions that can't supplement declining water supplies fast enough?' – which is an academic's way of saying, 'You think tensions are high in the Middle East now? You wait until the water runs out.'

The water problem alone should be enough to worry us, but there's more. Fertiliser consumption has risen fivefold since 1960 and overuse acidifies the soil, which has a *negative* effect on yields. Research by Fusuo Zhang (a soil and plant scientist at the China Agricultural University in Beijing) has shown that yields are down by as much as 50% in some areas, warning that if things carry on it could 'cripple Chinese agricultural production'. At the same time, chemicals in the fertiliser (notably nitrogen) find their way into local rivers, lakes and estuaries that, when they empty out into the sea, create huge marine 'dead zones'* – over 400 worldwide, covering nearly a quarter of a million square kilometres.

I'm sorry, but there's more.

The Green Revolution favours 'monoculture' crops – only growing one variety at a time as farmers seek 'economies of

---

* Algae in the water gorge on an excess bounty of fertilising chemicals and bloom, but when they die, the scavenger bacteria who munch on their corpses blossom too, consuming too much oxygen and suffocating other aquatic life. It's like the air has been sucked out of the water.

scale' – tailoring their irrigation, fertilisation and pesticide regime accordingly. But this lack of diversity in plants creates a similar lack of diversity in the insects, giving the pests that like to munch on your crops a free rein, because their own natural predators aren't sharing the field. The answer? Pesticides, which come with associated financial and environmental costs and the bugs often evolve around (in much the same way as tuberculosis has evolved around our medicines) – the so-called 'pesticide treadmill'. It's the same with weeds and their herbicides. The International Survey of Herbicide Resistant Weeds keeps an ever-growing list (471 as I write) of evolved weeds outwitting our chemical weapons. It's a decades-long arms race.

Then there's the increased soil erosion. Each year rain and wind erode about 70 billion tonnes of soil, a cavalcade of tiny particles that mostly make their way to the sea. This is a natural process, and it used to be that rotting plants (as they slowly break down into basic soil nutrients) would more than compensate for the losses. But our intensive farming techniques have reversed the trend. By ploughing the soil regularly, soil is broken up into finer particles (often to a greater depth than in the past thanks to modern machinery), which are more easily whipped up and washed away, while at the same time pesticides and fertilisers adversely affect the ability of the soil to bind (while also killing many organisms that help give it structure).

All this reduces soil fertility, encouraging farmers to increase Green Revolution inputs to compensate, which in turn accelerates the erosion – another vicious cycle.* On top of all that, much of the carbon that used to stay in the earth has found

---

* US soil erosion is estimated at 17 tonnes per hectare per year, with an associated *annual* cost of $44 billion.

its way into the atmosphere. Estimates vary, but between a third and a half of atmospheric carbon dioxide (CO_2) has come from our soils, a huge contributor to climate change.

It's all extremely bad news. As the old Chinese proverb goes, 'Man, despite his artistic pretensions and many accomplishments, owes his existence to a six-inch layer of topsoil and the fact that it rains.' Right now the very fields that ancient Chinese sage may have walked in, turned to the Green Revolution, are likely losing their topsoil *because* it rains, as the weakened and chemical-dosed earth can no longer hold on to itself.

Humanity is faced with a Catch–22. If we want high yields, we need to embrace the treadmill of engineered seeds, thirsty irrigation, pesticides, herbicides and fertilisers, with all the problems and costs this entails; but, in doing so, we simultaneously hasten the coming collapse in world water and food supply.

It's why what I've come to see in Jharkhand might be so important, because it could be a major component in addressing the dilemma. And, coincidentally, we've another Norman to thank for bringing it to the world's attention.

We stop at the village of Chitto, a modest collection of stone houses and dirt streets, surrounded by rice fields, all rain fed. Yezdi and Sudhir have brought us here because this is a village in transition. Some farmers have embraced a new rice farming technique championed by Erika, Gaoussou and KGVK while others have kept faith with traditional approaches (referred to by Yezdi, perhaps uncharitably, as 'spray and pray'), which means you can see the results side-by-side, and the contrast is striking.

I take two photographs of Sudhir standing in neighbouring plots that are using the different techniques. On the 'traditional' side the rice plants come to just above his knees. In the plot next door they're well past his waist. Gaoussou asks me to compare two panicles (the grainy fronds with the actual rice in them), one taken from each field. The panicle from the taller plant has many more grain pods and proportionally far fewer empty ones. 'A lot less chaff here,' he explains (suddenly I understand the origin of the phrase 'sorting the wheat from the chaff'). Erica wants to know how the number of tillers compare. Gaoussou and I estimate it's about 26 to 7 in favour of the taller plants.

'Both these crops are fed only by the rain?' I ask.

'Yes,' says Yezdi. 'And no chemical fertilisers are used either. They have the same inputs.'

'Last year we did an analysis of different yields between the farmers using these two techniques,' Sudhir tells us.

'What was your sample size?' asks Erika immediately.

'268' – a figure that seems to satisfy her. (Yezdi tells me KGVK is 'manic obsessive about capturing and processing good data').

'What was the difference?' asks Erika.

'An increase of between 35 and 50%.'

'Impressive!' exclaims Gaoussou.

Erika turns to me. 'For only rain-fed crops that's a great result. In my experience that's a lot.'

Sudhir is smiling. He's clearly enjoying the endorsement of the foreign visitors, and he still has another ace up his sleeve.

We've attracted attention. An elderly gentleman in a tight-fitting black cap and sporting a luxuriant white beard is observing us with a look of some consternation. Yezdi is quick to engage in conversation and, on hearing we're part of a KGVK delegation, the man smiles and becomes immediately amenable. Back in the car Yezdi reveals our visitor to be a village elder who's decided to encourage every farmer in the village (of which he is one) to adopt the new approach, something he refers to as SRI, next year.

'He was sceptical at first,' says Yezdi. 'Generations of traditional rice cultivation methods are not easy to abandon, but you can't argue when you see the results side-by-side.'

In 1983 Norman Uphoff went to Madagascar, came across the teachings of a priest and converted – not from one religion to another, but from the world of social sciences (in which he was a Professor at Cornell University) to the world of rice farming. It wasn't an immediate conversion. Initially Uphoff was sceptical of the claims he came across. In fact it took three years of field data for him to associate his name, and that of his university with the idea. ('I was very conscious of the fact that I could not/ should not associate Cornell's name with anything that was not well-founded, anything that could turn out to be a hoax, or a figment of imagination,' he says).

Though Uphoff never ('to my great regret') met Father Henri de Laulanié, he did become intimate with the priest's work via his hosts in Madagascar, the NGO Tefy Saina. A Jesuit missionary based near the city of Antsirabe, de Laulanié had been working for many years with local farmers in efforts to improve their agricultural systems, paying particular attention to the nation's staple crop of rice. He created a rural training centre and it was here that an accidental discovery set him on the path to notoriety.

A standard technique in rice farming is to grow seedlings safe in the confines of a greenhouse for about 30 days, allowing them to establish themselves before transplantation to the field. However, a delay in seeding the Antsirabe nursery one year meant the baby rice plants had to be transplanted after only 15 days, half the usual time. Henri worried these younger seedlings would fare poorly, but instead they flourished, producing roughly double the numbers of tillers per plant than the normal average of 10. Intrigued, the priest experimented with even earlier transplantation, 'at only 12 days, 10 days, even 8 days'. To his surprise this led to 'a substantial increase in the number of tillers per plant, up to 60, even 80 or more'. Given that each tiller is usually the forerunner of a panicle, these high numbers represented a massive upswing in productivity. The harvest was booming.

The only problem was Henri had no idea why. It took four years and the publication of Didier Moreau's gripping thriller, 'L'analyse de l'élaboration du rendement du riz: les outils du diagnostic' (Diagnostic Tools for the Analysis of Rice Yields) to find out. In it de Laulanié found a discussion of Japanese researcher Tsukuda Katyama's work investigating how wheat, maize and rice plants develop over time. Katyama discussed plant 'phyllochrons' – the period between a tiller growing and it budding its own 'child' tillers. Katyama had found that the later you moved a seedling from the nursery to the field, the longer it took it to recover from the trauma, reducing the number of phyllochrons ('tillering' opportunities) it could get through before maturity. Seedlings transplanted earlier got back on track much faster, and enjoyed more 'tillering periods' as a consequence.

De Laulanié continued experimenting relentlessly in his own fields, observing local farmers' different approaches and reading all the academic literature on rice farming he could get his hands on (often slim pickings in rural Madagascar). Over time he developed recommendations on how and when to weed your plots, when to wet the soil and when to let it dry, the optimum times to harvest and how and where to place seedlings. One of the techniques central to SRI is a wider and even spacing of plants, the theory being that this creates less competition between the roots of different plants for soil nutrients and more sunlight can reach more of the leaves, promoting growth.* Indeed, one of the first things I noticed about the SRI plot in Chitto was the regular spacing of the plants, as opposed to the more haphazard arrangement of the field next door.

'That obviously takes more effort during planting,' says Erika. 'But you saw each plant had many more tillers than in the traditional approach.' Helping farmers achieve the even spacing with less effort is another obsession of Erika's. At the KGVK campus last night she'd pored over a hand-pushed seeding machine designed to place seeds in the regular pattern SRI suggests, asking Sudhir and Yezdi a barrage of questions about how they were finding it.

De Laulanié dubbed his collection of techniques the 'System of Rice Intensification based on Katayama's Tillering Model' or 'SRI' for short, but stressed his recommendations should be adapted depending on each farmer's individual circumstance, writing, 'Observe and listen to your rice. It is the only one who can really tell you what should be done.'

And that might well have been that, if it hadn't been for Norman Uphoff who, inspired by bumper harvests achieved without massive inputs of fertiliser or irrigation, dedicated the rest of his professional life to promoting SRI, convinced that the priest's work offered a great hope for mankind, allowing us to maintain Green Revolution yields without the sustainability headache we've inherited alongside. He founded Cornell University's SRI

---

* There is, I find out, a rich vein of SRI literature concerning rice plant spacing with varying advice on layout depending on climate, the type of soil and so forth – although the most common recommendation is transplanting seedlings into a square grid with 25cm between each.

International Network and Resources Center, which remains the pre-eminent hub for SRI knowledge-sharing, and recruited Erika to head it up when he retired.

Not everyone was happy. As the news of SRI began to spread, the reaction from established rice scientists grew from dismissal to withering disdain. They didn't take kindly to a French priest and a social sciences professor getting in on their act. While Uphoff called only for a broadening of rice research to more readily embrace 'whole system' approaches, many interpreted his work as full-on criticism and an unwelcome challenge to years of existing Green Revolution research (and, by association, the scientists who conducted it) – and, worse still, coming from a bunch of amateurs. A particularly vitriolic exchange of views between 2002 and 2004 even became known as 'the rice wars'.

'You couldn't believe some of the things that were written,' Erika tells me the next day as we drive out for a series of field visits. 'I mean, it was unbelievable. It wasn't proper debate.'

Dr Thomas Sinclair, a professor in Crop Science at North Carolina State University, went so far as to suggest that 'discussion of the System of Rice Intensification is unfortunate because it implies SRI merits serious consideration,' which is the academic equivalent of saying 'talk to the hand'. In his opinion, SRI is a non-idea and even discussing it is giving credit it doesn't deserve, the legitimisation of nonsense. He described the evidence offered as 'Unconfirmed Field Observations' or 'agronomic UFOs', about as credible as 'their space UFO cousins'.

Sinclair's wasn't the only cold shoulder experienced by Uphoff. When he sent a draft paper on SRI to the International Rice Research Institute (*the* rice research body set up as the Green Revolution got going and dedicated to developing 'advanced rice varieties that yield more grain and better withstand pests and disease'), the response was that five of the six basic SRI practices had already been investigated by rice scientists elsewhere and so there was really very little of worth to comment on. In other words, de Laulanié and Uphoff were re-inventing the wheel, and not very well. But neither claimed they were proposing something entirely new. De Laulanié was clear he derived a lot of his system from the work of others. What they were arguing for was a *synthesis* of techniques that together could deliver more than the sum of their parts.

It didn't help that SRI advocates started reporting per hectare rice yields of over 15 (and sometimes 20) tonnes per hectare, which to mainstream rice scientists marked them out as nut jobs. Rice yields, averaged worldwide, hover between 4 and 5 tonnes. Claims of three or four times this were dismissed as 'probably a consequence of measurement error', because such yields would, it was claimed, exceed rice's 'photosynthetic efficiency' (the amount of the sun's energy a rice plant is able to turn into growing power).

And yet the claims keep coming.

In 2012, Sumant Kumar, a farmer using SRI-inspired methods in Nalanda (a poor province in Jharkhand's neighbouring state of Bihar) claimed a yield of 22.4 tonnes per hectare, a world record. The harvest was reportedly verified by the local government and Sumant became something of a celebrity. The *Observer* reported how 'The state's chief minister came to congratulate him, and the village was rewarded with electric power, a bank and a new concrete bridge.' Ironically, Sumant's new-found status as rice hero meant he had less time to attend to his crop, and the following year his yield suffered. That said, he still managed 13.5 tonnes per hectare, a threefold advance on the average harvest. So, using SRI, when he wasn't paying much attention, Sumant was still achieving Green Revolution yields without the expensive inputs or negative side effects.

Or was he? A lot of the reporting at the time brushed over the fact that Sumant wasn't working alone, but collaborating closely with the Nalandan branch of India's Agricultural Technology Management Agency (ATMA). Reporters love a 'lone genius' story, even if it's not true. With the help of ATMA, Sumant was mixing and matching techniques and tools. Yes, he used a largely SRI approach, but he was also using fungicide-treated hybrid seeds from the agri-giant Bayer. Neither did he entirely shun chemical fertilisers or herbicides, using small amounts where he thought it appropriate. Breathless second-hand reporting claiming 'Indian farmers smash crop yield records without GMOs' or 'chemicals of any kind' were flat out wrong. All that said, Sumant's inputs *were* much lower than a full-on Green Revolution approach, *particularly* when it came to water usage.* De Laulanié would no doubt have approved – a farmer combining techniques and technologies in response to his own situation and the best knowledge available to him, a farmer engaged in active research while 'listening to his rice'.

There are no strict edicts that state you must use every technique in SRI's portfolio, that you can't mix and match with other approaches if it suits you. Proponents of SRI point to superior yields like those of Sumant Kumar as proof the system clearly works wonders. If they're diligent, they'll also acknowledge his integration of non-SRI components as exactly the sort of local adaptation they advocate. Critics flip the argument around, saying this is essentially a cop-out – it's not SRI.

My own analysis is that de Laulanié made an unwitting slip-up when he came up with his title 'System of Rice Intensification'. Once you call something a 'system' it implies predictable inputs

---

* It should be noted that some sceptics still feel that Sumant's record-breaking yield can't be true. Yuan Longping, the Green Revolution rice scientist, who believes he still holds the world record at 19.4 tonnes, called Kumar '120 per cent fake.'

and outputs – a machine, perhaps complex, but where you understand that if you put certain things in one end you'll get foreseeable and repeatable results out the other. Indeed the systemisation of agriculture is one of the hallmarks of the Green Revolution, seeking to remove uncertainties from the field through the use of monocultures, pesticides, fertilisers and irrigation schemes. De Laulanié's approach actually isn't a 'system' at all, but a kitbag of tools that might help you work more productively with what you've got. It has therefore increasingly found favour with smaller farmers who can't afford to transition to all the expensive inputs that the Green Revolution requires, explaining much of its popularity here in Jharkhand, where Yezdi tells me '90% of farmers are smallholders with less than a hectare, mostly growing cereals or pulses for subsistence. They can't afford a cock-up.'

I suspect that if de Laulanié had chosen the title: 'Toolbox of Techniques for Rice Intensification', a lot of the squabbling over its claimed merits would have been averted. Deputy Director for research at the International Rice Research Institute, Achim Dobermann, is often characterised as a critic of SRI. His view is that 'SRI is a set of management practices and nothing else, many of which have been known for a long time and are best recommended practice.' On Sumant Kumar and his friend's mega-yields (of which Dobermann disputes the size but doesn't doubt was an excellent crop for which they should be congratulated),* he says, 'Scientifically speaking I don't believe there is any miracle.' The irony is that de Laulanié would almost certainly have agreed with him.

'All we're saying is you have to start with where the farmer *is*,' says Erika as we head to our next field visit, echoing de Laulanié's central principle. Those words come to have a particular

---

* Dobermann estimates the 'yield may have been more like 10–12 tonnes a hectare'.

resonance when Sudhir, at the end of our trip, reveals his *pièce de résistance*, something Erica and Gaoussou find truly astonishing.

For the next two days we visit rice field after rice field, chatting to farmers and local communities. In a village high in the hills, Erika, Gaoussou and I are welcomed as honoured guests and adorned with huge flower necklaces. Via Sudhir and Yezdi, residents tell us how increased rice yields, with the help of KGVK, are translating into better healthcare and education, as the economy of the village improves. Gaoussou recounts a story of how, in his native Mali, the extra income that came to one male farmer using SRI encouraged him to take extra wives (polygamy being common and accepted in Mali).

'So it became a system of *wife* intensification for him, then?' I say, and he laughs uproariously. (I can tell you that having Dr Traore laugh at one of your jokes, however weak, makes you feel like you've sold out the Hammersmith Apollo.)

'It's not the message we want to promote!' he replies. It turns out Gaoussou is missing his own wife (singular), who is clearly the one true love of his life.

Everywhere we go, Erika is checking to see which of the core SRI principles are in use. What kind of fertiliser, if any, is being used? When are the seedlings transplanted? How are they spaced? Are they transplanted singly or in groups? (SRI recommends one plant per spot on the grid to avoid root competition.) What's the weeding protocol? (In tandem with the early transplantation, de Laulanié proposed a rigorous weeding regime so that his youthful transplants wouldn't be overwhelmed by leafy aggressors.)

Another village, and again we're treated like VIPs, invited to observe a traditional dance before Erika is asked to make a short speech about what she's seen in the surrounding fields and to offer any advice. In a larger village down in the valley, a co-operative of women farmers proudly show us their plots, and tell us how SRI techniques have improved their lives. We hear stories of yields doubling, of rain-fed plots planted with local seeds delivering harvests of 10 tonnes or more per hectare, which even Erika finds hard to believe ('That is too much, that's really, really high').

It turns out I've been extraordinarily lucky with my timing, arriving two weeks before the harvest, meaning the fields are at their most bounteous and easiest to compare. Time and time again we see neighbouring fields where traditional 'spray and pray' methods are sat right next to SRI-planted plots – and the comparisons are hard to argue with. Here in Jharkhand, de Laulanié's 'set of management practices' *are* delivering Green Revolution yields in rain-irrigated fields without an accompanying sustainability headache. Erika regularly expresses her admiration, which, given she spends most of her life travelling the world to visit SRI projects, means that something is going very right here. Every time she does, I see the young Sudhir Paswan, KGVK's resident miracle worker, quietly smiling to himself.

Not yet 30, Sudhir Paswan is KGVK's head of Agriculture and Farmer Education. In short, he's *the* local font of knowledge on how to improve yields and the fields we've visited these last few days are testament to his good works. But now he wants to show us one farm in particular: his own.

As we get out of the car, Sudhir tells us, 'Before I took over this operation things were very bad. To be honest, the feeling was you couldn't grow anything much here. People thought I was crazy.'

'Because?'

'Because it's an upland farm, we're far from the water table. The soil is bad too. You wouldn't normally farm an area like this unless you didn't have another option.'

Gaoussou, with his expert eye, concurs. Apparently we're standing on 'laterite' soil, rich in iron and aluminium and with a high clay content that can easily develop a hard crust, greatly hampering its ability to hold water. As we walk, it's hard for me to believe the farm's location can be as bad as Sudhir is making out. We're surrounded by verdant abundance. Two women pass us carrying huge bags of just-harvested peanuts. Sudhir invites us to nibble on a few as we head uphill, passing lines of fruit and vegetables planted according to SRI techniques. De Laulanié's toolbox now enjoys a wider stage than rice cultivation. On our first evening at KGVK's central campus, where Erika had been wowed by gigantic rice plants, we'd also seen a host of other crops, including sweetcorn and gherkins, planted using SRI-like techniques. That's why in many places SRI is now referred to as the 'system of *root* intensification' and, again, the results for other crops can be as astonishing. Sumant Kumar, who attracted all those headlines with a reported rice yield of 22.4 tonnes per hectare, was soon followed by his friend Nitish, who smashed another world record using SRI techniques – but for potato yields. He achieved 73 tonnes per hectare (the previous record being 45). But even he was eclipsed the following year by Rakesh, chairman of the local organic vegetable-growers federation, who, again taking SRI as his inspiration, harvested 109 tonnes per hectare.

There are, of course, rice plots here too. Sudhir is testing different seed types and spacing. Erika, as ever, is quizzing him, particularly on the fertiliser he uses on what she agrees is highly unpromising ground. Sudhir points to a hut behind and below us where he mixes cow dung and urine in careful ratios (surely one of the world's least palatable jobs) and she notes down the formula.

'This is very unusual,' Erika tells me. 'You wouldn't normally see rice on an upland farm with soils like this, especially in an area with such erratic rainfall.'

'Rice needs moisture,' interjects Gaoussou. 'On a lowland farm, even if it's only rain-fed you'll usually get some flooding during the season. That means there's more water seeping into the soil. Rice is also one of the few plants that can survive being in standing water, so it makes sense to plant it in lowland areas if you can. Not here.'

We've reached the topmost plot of the farm.

'My goodness!' says Gaoussou. '*Here?*'

'If you dig one foot down you will find hard rock,' Sudhir tells us, a broad grin on his face.

Erika bends down to examine the earth. 'This is *really bad* soil. You would never think of growing in soil like this. Never!'

And yet we're standing next to a plot of healthy-looking rice plants growing away, happily unaware of their celebrity status.

'Why did you do it?' she asks.

'To see if we could,' replies Sudhir. 'To do the impossible. If we can grow rice here, on top of this rocky hill with only a little bad soil, fed only by the rain, using SRI ... It's a powerful teaching tool. I show farmers this, they pay attention.' He turns to Erika. 'But Doctor, do you have any advice on how we can improve things?'

She snorts. 'Advice? You've got to be kidding me? I should be asking *you.*'

Sudhir Paswan gives me enormous hope. He's shown me that there are alternatives to our modern agricultural system. If farmers here in mostly rain-irrigated Jharkhand can boost their yield by up to double with little reliance on chemical fertilisers and without plundering the water table, it shows there *is* a way to overcome the dilemma presented by the current version of the Green Revolution. Increased yields needn't come tied to the eventual destruction of the very foundations of our agriculture (in particular our water sources). The mantra that only massive agri-business can feed the world simply isn't true. Here I am, standing on a plot of land that apparently only a madman would think to farm on, and yet it's blooming. Sudhir, and all the other farmers I've met, have shown, in the most direct way possible, that our food future need not be so bleak.

SRI is one example of what is known as agroecology – designing agricultural processes that work *within* an existing ecosystem, rather than trying to replace it, or keep it at bay. Because of this, agroecology is available to all farmers, even the poorest, which is good news. It doesn't insist they invest in expensive fertilisers or irrigation systems. Here in Jharkhand, I've heard story after story of how an agroecological approach has helped transform the fortunes and aspirations of small farmers and their families.

The United Nations Food and Agriculture Organisation estimates there are 500 million family farms (the vast majority under 2 hectares) worldwide. Collectively they are responsible for 56% of global agricultural production. Currently, many are run on a subsistence basis, which might suggest a romantic image of honest rural folk in tune with their environment, but the clue is in the name. Subsistence, by definition, is 'the action or fact of maintaining or supporting oneself at a minimum level'. For this reason, and for many years, family farmers have been seen as part of the problem of hunger, bolstering the argument that we need increasing industrialisation of farming.

And yet Professor Olivier De Schutter, former special rapporteur on the right to food for the United Nations, concluded that family farmers could double food production within a decade using agroecological methods like SRI. His 2011 report to the UN brought together research from 57 developing countries, showing an average crop yield increase of 80% for farmers embracing agroecology. That's one hell of a statistic, which you'll not hear the agribusiness lobby quoting very often. It's a statistic that could change the world.

'Conventional farming relies on expensive inputs, fuels climate change and is not resilient to climatic shocks. It simply is not the best choice anymore,' wrote De Schutter. 'We won't solve hunger and stop climate change with industrial farming on large plantations. The solution lies in supporting small-scale farmers' knowledge and experimentation, and in raising incomes of smallholders so as to contribute to rural development.' De Schutter is asking for a worldwide echo of Father Henri de Laulanié's approach. It turns out small farmers aren't part of the problem at all, but most of the solution.

It's my final evening in Jharkhand, before I head to the UK to investigate the almost unbelievable story of a man in a shed inventing a seemingly impossible machine that could be another ingredient in a much-needed food system reboot. Our group is gathered at the faded but charming Ranchi Gymkhana Club, where, over an excellent curry, I mention De Schutter's work and an observation he makes: that, despite the growing body of evidence, agroecology is still treated by many as a fringe idea.

'Ego is the biggest problem,' says Sudhir. 'If I go to a rice conference and tell them we're getting 14.3 tonnes a hectare, they don't believe it, they say I am making fake data.'

'SRI was developed for small farmers by small farmers. It wasn't created in a lab,' says Erika. 'The farmers *know*, but if you ask the scientist to leave his lab and talk to the farmer as *the expert*, that's a big knock to your prestige.'

Echoes of my previous two stops are sounding in my head. In Boston I learnt how patients could outperform existing clinical research methods at a fraction of the cost using repurposed Internet dating technology. In Delhi, crowdsourced student researchers bypassed the experts to create the world's best-annotated tuberculosis genome. Here in Jharkhand, rural farmers are exceeding the yields of the Green Revolution with a bag of techniques refined from the work of a Jesuit priest.

'If some rice scientist came up with a seed that could do what Sudhir's doing they'd be getting a Nobel prize!' says Gaoussou. (What was it PatientsLikeMe CEO Ben Heywood had said about the improvement they've seen in epilepsy patient's outcomes? 'If I could create a *drug* that could do that, I'd be a very rich man.')

'Remember that rice plant we saw on our first night here?' says Erika. 'I mean, that thing was like a *bush!*'

'SRI makes money for poor farmers, not big corporations,' says Sudhir.

'How many farmers are using SRI-like techniques?' I ask.

'It's hard to say,' replies Erika, 'because, as you know, there's such a debate as to what is SRI and what isn't. But we did a census at the end of 2013 that provided us with some partial data and estimates. Best guess? About 10 million farmers in 57 countries on between 3 and 4 million hectares of land are benefiting. That's in *all* environments, from the desert regions to humid tropics, for rain-fed or irrigated rice, as well as other crops.'

'So, at best, two per cent of the world's family farms?'

'Yes, but the momentum is growing. You've seen the results here, side-by-side. First one farmer goes, then his neighbours, then the whole village.'

I think back to the village elder on our visit to Chitto, and his desire that every farmer there should embrace SRI next year.

'The results. I've seen them, and I've seen them all over the world,' says Erika. 'What the farmers are achieving here in Jharkhand? It's amazing, isn't it?'

She's looking straight at me.

'It is amazing,' I say, and it's not out of politeness. I'll admit it, I'm excited by rice.

5 RUNNING ON AIR

'The person who follows the crowd will usually go no further than the crowd. The person who walks alone is likely to find himself in places no one has ever seen before.'

– ALBERT EINSTEIN, SCIENTIST

I'm sitting in a London restaurant with a man who, by his own admission, attracts nutters. 'Pretty much every time I go on TV, the next morning my inbox is full of them,' he sighs. 'Perpetual-motion machines, faster-than-light engines, time travel…'

Imagine Tintin in his mid-fifties and you've pretty much got Dr Tim Fox, a man often called upon by the media to pass comment on stories with an engineering flavour. It's part of his role as Head of Energy and Environment for the UK's Institution of Mechanical Engineers which makes him an easy-to-find target for crackpot inventors the world over. So when Tim was called to meet two men peddling a radical advance in an age-old technology, he wasn't exactly jumping up and down with excitement.

Tim entered the wood-panelled George Stephenson room at the Institution's London HQ and, underneath a painting of the famous man and his Rocket steam engine (something of an omen of things to come), was introduced to Toby Peters and Gareth Brett. 'I'm thinking "OK, here we go, then, another couple of crackpots, ten minutes of my time, I'll be polite and then I'm off." My face was a screensaver.'

Three minutes later his cynicism lay in tatters.

'It was just one of those moments you live for,' he says, smiling. 'It changed everything.'

It started in a shed. Indeed, the shed is still there, in Peter Dearman's backyard, which you can find in the old English market town of Bishop's Stortford, Hertfordshire. Through its windows you will often see Mr. Dearman, probably with a spanner in his hand. He's the archetypal shed-hobbyist: jeans, anorak and totally obsessed with mechanical apparatus. His neighbours jokingly used to call him 'the neighbour from hell' thanks to the discarded bits of machinery that lay about his property. Today they're slightly more respectful. As they should be – because, early in 2000, Peter solved a problem whose solution had evaded the engineering world for a century. In doing so, he unleashed a coming revolution that has the potential to save millions of lives. And he did it with a can of antifreeze.

Peter's fascination with all things engineering began as a child. The third of nine siblings, he and his older brother John would 'do lots of different scientific experiments' around their parents' poultry farm. 'There was the occasional bang but we never really blew anything up – although that was more luck than diligence,' he admits. When Peter was eleven, John died in a car accident, leaving his younger brother bereft; but within a year, he continued his tinkering, becoming increasingly interested with the problem of resource shortages. 'I started to think, "How would the world function without oil? How would we get anywhere?" I knew it had to run out sometime and I wanted a solution for when it did.' Not yet a teenager, he started to investigate alternatives to the oil economy – a quest that would eventually lead to him solving an entirely different problem.

First he turned to electric vehicles. These were few and far between in the 1960s, although the UK did have a fleet of electric-powered milk floats. Could their battery power be one option for replacing the petrol engine? Peter soon concluded the answer was 'no'. Batteries simply weren't 'energy-dense enough' – an engineering term for how much oomph can be released from an amount of matter. (This was long before electric and 'hybrid' cars began to enter the mainstream. Elon Musk, CEO of Tesla, wasn't even born when Peter started his investigations.) In fact, the energy density of fuels derived from oil is one of the key reasons they continue to dominate the world – they pack a mighty punch in a very small package. To give you an idea, gasoline is roughly twenty-six times more energy-dense than the consumer batteries you put in a hand torch. Oil company execs like to quote statistics like 'one gallon of gasoline is enough to charge an iPhone once a day for almost twenty years'. The only fuels we've harnessed commercially that are more energy-dense are thorium and uranium. A chunk of uranium taking up roughly the same space as a gallon of petrol would charge your iPhone every day for over forty-five million years.

Undeterred, the young Peter Dearman had a look at domestic storage heaters, which, rather than storing energy as electricity, store it as heat. He built his own prototype, warming up a block of concrete, and using the amassed heat to power a small steam engine. His storage heater still fared miserably in terms of energy density compared to petrol, but in building it the thought occurred 'that you didn't have to store energy only as heat. I thought, "I could use cold. Why don't I try storing energy as *cold?*" '

It's not an idea that immediately makes sense to most of us. When we think of energy we tend to think of hot things: fire, steam, warm computer batteries ... Cold is something we associate with a *lack* of energy. But our intuition is misleading. Two objects of different temperatures brought into contact will exchange energy (as they try to get to the same temperature) even if those temperatures are both below freezing point – and if you piggy-back on that exchange you can do useful work. Peter, of course, wasn't the first person to realise this, but he did become the first person, many years later, to use the concept to create a machine that could help to save millions of lives by changing the way our food system works – 'The Dearman Engine'.

My brother-in-law Mark Smith is a brilliant mechanic, one of those practical types who make you feel slightly unworthy every time you look in your toolbox. He tells me, 'When it comes to engines, you need to learn the basics and the basics haven't changed for a long time.' In fact, engine design hasn't altered significantly for over two centuries ('most of what we've got today is enhancements that add a little bit of efficiency here, or some more power there – but it's the same beast underneath,' says Mark). Perhaps it's this very lack of change that explains why most of us don't really know how engines operate, in line with Douglas Adams's observation that 'anything that is in the world when you're born is normal and ordinary and is just a natural part of the way the world works'. Engines just *are*. In fact, a quick survey of my local pub reveals that, when it comes to their operation, most of us know pistons are involved somewhere along the line, but that's about it. How those pistons are set in motion, or how their movement is converted into useful work, like turning a wheel, we're generally hazy about.

So, some history. The first steam engines were built over two thousand years ago and called either 'aeolipiles' or 'Hero engines' (after their suspected inventor, the fabulously named Hero of Alexandria). The first engine took the form of a metal sphere filled with water that, when heated to boiling point, sent steam out of opposite facing 'jets' on either side of the device, causing it to spin on its axis. Today, we're more likely to associate the words 'steam engine' with classic locomotives like the historic *Flying Scotsman*, vehicles which provide as good a model as any to understand the basics of how engines work.

A heat source (in a steam locomotive that's the big fire the driver and their mate keep shovelling coal into) warms water in the boiler (the big, distinctive horizontal cylinder at the front of the train that forms the bulk of the engine). So heated, the water boils, creating steam – steam that can be guided through a thin pipe. This piped steam is channelled into one side of a piston chamber. Freed from the constraints of the narrow tube it just came through, it expands into that chamber, pushing a piston (one 'stroke'). The valve closes and another opens on the *other* side of the piston chamber, where more fresh steam duly expands and pushes the piston *back* where it came from (the reverse stroke). Back and forth, back and forth. The sounds that accompany this ballet of mechanics give rise to the distinctive chugging and hissing sounds we associate with steam locomotives. Pulleys and gears can transform the back-and-forth motion of the piston into rotation, driving the wheels.

Dearman Engines work on exactly the same principles, except they're powered not by steam (water gas) but by another gas, which, when you first hear about it, seems to make no sense at all. The gas that drives a Dearman Engine is literally thin air.

I've come to the Kensington campus of London's prestigious Imperial College to meet Peter. Now in his mid-sixties (though he looks considerably younger), he immediately strikes me as a gentle soul. If you stuck him in a dog collar he'd make the identikit kindly English vicar.

We're camped in a small office, having just visited a bijou basement laboratory where two newly-minted prototype Dearman Engines have been successfully put through their paces. I have to confess I'd been slightly disappointed. I'd rather hoped to feast my eyes upon a new category of machine, some kind of steampunk/sci-fi crossover, but to my untrained eye Peter's machines looked unremarkable. The assembled engineers (I counted six crammed into a space not much bigger than a small studio flat), however, all seemed very happy. Everything is going well, they told me. It's not what the engines look like that counts. It's what they do.

As a stimulus for our chat I've brought a photo, dating from the turn of the twentieth century, showing a couple riding a rather flimsy-looking car. The caption describes a 'graceful little motor-car' reportedly powered by air that 'makes absolutely no smell or noise'. The picture is credited to a company founded by Danish engineer Hans Knudsen. It's doubtful, however, that the car in the photograph was powered by air. More likely it's an early example of what people in advertising might call 'aspirational marketing' – and you and I would call 'a lie'. Although Knudsen's company, set up to manufacture air-powered cars in Cambridge, Massachusetts, in 1899, gained much media interest, 'when reporters wanted to see the factory, they were given the run-around from one location to the other. The only evidence produced was a drawing of the car.'

That's not to say Knudsen was a charlatan. The company made serious attempts to create an air-powered automobile, setting up factories in the UK and the USA. Novel ideas in motoring abounded in the late nineteenth and early twentieth centuries and competed fiercely. *Most* automotive manufacturers failed, or became absorbed into the few big names that prevail today. Knudsen's inability to produce a working prototype was almost certainly down to the fact he never had the insight Peter Dearman was to have over a hundred years later, although he had got as far as converting the air into the right state for use as an energy source – that is, a liquid.

It's hard to think of air as a liquid, but like most substances, air can exist as a gas, a liquid or a solid, if you compress and cool it enough. In fact, air becomes a liquid at about minus 195°C and freezes solid at roughly minus 215°C.

What Knudsen didn't know about making liquid air wasn't worth knowing. In his brilliant and entertaining book *The Romance of Modern Invention,** the Edwardian science writer Archibald Williams describes his 1902 visit to Hans Knudsen's factory on London's Gillingham Street, where he was shown around by the man himself. Through repeated rounds of compression (getting to 'a ton of pressure on the area of a penny'), cooling and a smidgen of low-temperature evaporation, Williams described how the molecules in the processed air 'utterly deprived of their self-assertiveness' collapsed into 'a clear, bluish liquid, which is the air we breathe in a fresh guise'.

---

* Full (awesome) title: *The Romance of Modern Invention Containing Interesting Descriptions in Non-technical Language of Wireless Telegraphy, Liquid Air, Modern Artillery, Submarines, Dirigible Torpedoes, Solar Motors, Airships, &c. &c.*

Liquefied air is 700 times denser than that we inhale. What's more, it will return to its gaseous state almost instantly if you open a valve on a tank of the stuff. And if you want to push a piston, something that can quickly expand 700fold in volume can come in handy. As a young boy Peter Dearman investigated liquid air-powered engines in his quest to replace oil, but soon found that the laws of physics were stacked against any such endeavour. Even compressed to 1/700th its original size, air has far less energy density than a fossil fuel.

'So I put it on the back burner,' he says. 'It was just something I would think of occasionally.'

Besides, he had other things to occupy his attention – like getting thrown out of school. Not because he was a poor student, but because he was too good. One of Peter's teachers set an exam including a question about converting temperatures from Celsius to Fahrenheit – and she'd handily provided the formula needed. Except it was wrong.

'She'd put one of the fractions upside-down. There was no way you could get the right answers using it' – although Peter did get three out of the four right through some calculated guesswork. When the teacher went through the exam with the class, she fudged her workings to hide her mistake. 'So, of course, I stood up in the class and told her,' says Peter. 'And that was the end of me!' Rather than admit her error, Peter tells the story of how she accused him of cheating and by mutual agreement he left the premises, aged fifteen, to work in a sheet metal factory.

Perhaps one of the most promising engineering minds of his generation, Peter Dearman does not possess a single qualification and was run out of school thanks to the false pride of teaching staff. Despite this, he's enjoyed a long career as an engineer, working in various fields, but always in his shed in the evenings and at weekends. At one point he invented a new 'automatic cycling circuit' for medical resuscitators ('you know, kiss of life machines') that was much more efficient than its predecessors – essentially allowing the devices to run more cheaply and be considerably less bulky. That wasn't a smooth ride, either. The sales and marketing person Peter worked with to get manufacturers interested later tried to claim the invention was his – and took Peter to court. The judge found in Peter's favour, referring to him as 'a good, honest witness' and 'essentially an engineer' who was largely uninterested in matters of business.

Then, late in 1999, Peter was watching one of the final episodes of *Tomorrow's World*, the BBC's prime-time science magazine programme. In it presenter Peter Snow visited the University of Washington to report on a liquid-nitrogen-powered vehicle built into the body of an old mail truck. (Air is a mite over 78% nitrogen, meaning liquid nitrogen is as close to a liquid air as diesel is to petrol.) Even with the presenter cooing over it, the researchers had to admit the vehicle had major shortcomings. It was 'fabulously inefficient', they acknowledged, and took 'gas guzzling' to new heights 'by consuming about five gallons of nitrogen fuel per mile'. The truck's top speed? An underwhelming 35 kilometres per hour. And you didn't ask how it handled hills.

'This motor operates at less than twenty per cent of the efficiency we think is possible,' wrote John Williams, a graduate student working on the project. 'We know we can do better' – except they didn't really know *how*. The Washington researchers had hit the same fundamental problem that had sent Hans Knudsen's Liquid Air, Power and Automobile Company bankrupt. Air, even in liquid form, simply didn't have enough oomph as a fuel. It was all down to energy-density again. While a gallon of gasoline can charge your iPhone every day for twenty years, a gallon of liquid air could only do so for about five months. But even with these inefficiencies, the Washington team thought the engine had some merits, most notably that it operated at the *same price per mile* as gasoline and it ran completely clean.

Watching TV in Bishop's Stortford, Peter Dearman thought back to his youthful investigations into air engines and had a brainwave. He bought his first Internet-connected computer so he could download the Washington research. In those papers he found a statement of what was keeping the Washington engine from achieving higher efficiencies (which chimed with his own analysis) and an idea of how to overcome this, which was arguably laughable – requiring an enormously complicated redesign and submerging the entire engine in water. Peter had already imagined a much simpler solution.

'The rate at which a gas expands is down to the amount you heat it,' he explains. More heat means faster expansion because you're giving the molecules in the gas extra energy. They'll bounce around more and therefore push harder against anything they hit (e.g. a piston). This is where fossil fuel-powered engines have an ace up their sleeve. Gasoline gets turned into that energetic gas by being ignited and vaporised (that's what your spark plugs are for). The liquid fuel goes almost instantaneously from the temperature in the tank to 700°C or thereabouts. That's a lot of heat, which translates into a rapidly expanding gas.

'With liquid air we're going from minus 200°C to whatever the temperature in the engine is,' says Peter. 'But that's less than a 300°C difference. In a petrol engine you've got temperature change that's over double that. And that's only half the problem.'

'What's the other half?'

'Things cool fast when they expand. What power you do have drops off really quickly.' The expanding gas pushes the piston away, but in doing so increases space in the chamber. Pressure and temperature drop off rapidly, halving and halving again. The petrol engine gets over this problem thanks to the energy density of the fuel and the initial ignition giving the expanding petrol-gas an almighty heat-kick at the get-go. The fuel-gas is actually on fire, keeping it excited and pushing pretty hard against the receding piston.

Peter realised an air engine would never fully compete with its fossil fuel counterpart on power, but he knew he could raise its power to useful levels – creating a cheap-to-run and completely clean engine that could do useful work. 'What was needed was *isothermal* expansion,' he says, forgetting perhaps that I don't have a degree in thermodynamics. However, as a writer I know my prefixes. 'Iso' comes from the Greek *isos*, meaning 'equal' or 'uniform'. 'You mean, keeping the air in the piston at the same temperature even as it expands?' I ask tentatively.

'Yes,' replies Peter. 'If you can keep the temperature up, the expanding air will keep pushing longer, as opposed to cooling down and the pressure dropping off. The engine will have more power.'

The irony for liquid air engines, until Peter's intervention, was that a heat source to keep the air expanding at a constant rate (that 'isothermal expansion') was only millimetres away, but inaccessible. Given the air was starting off so cold (liquid air is minus 195°C, remember) the ambient temperature around the engine would be enough to keep exciting it, if only it could find its way into the piston chamber. The problem the Washington team (and everyone before them) had hit was they had no workable solutions for getting that ambient heat channelled into the piston chamber quickly enough. Within the chamber's confines, everything got way too cold way too soon, condemning the engine to low efficiencies. Each stroke simply ran out of puff.

But, once he'd realised 'isothermal expansion' was the missing link, to Peter the solution seemed obvious. Except, of course, nobody had thought of it before.

Peter wandered into his shed and hacked his lawnmower to run on liquid air, but with one crucial innovation. He injected antifreeze (a liquid specifically designed to be a particularly efficient heat exchanger) into the piston chamber on each stroke. Now heat from outside the piston chamber, carried in the antifreeze, had a far quicker route into it, delivering the isothermal expansion Peter needed. The efficiency of the engine leapt. Although he didn't know it at the time, Peter Dearman, with a keen mind, a lawnmower and a can of antifreeze, had just changed the world.

Like his Washington predecessors, Peter thought there was probably *something* in it. By his calculations, a full-size version of the engine would now be *cheaper* per mile to run than a gasoline equivalent, running on a completely clean and abundant fuel – although neither of these characteristics would be the thing that now has the world beating a path to his shed door.

Initially he was reluctant to tell many people. 'Being publicised is the horror of my life, the one thing I never want to do,' he says. 'I'm not the sort of person who goes around persuading people. I'm happier simply showing them what I can do.'*

Peter's 'show don't tell' attitude led him to hack his car to run on liquid air, but it took another brother, Andrew (a builder – they're a practical family the Dearmans), to move things to the next stage. Andrew was working on the house of recently retired energy industry executive Ferdinand Berger and mentioned to his client that his brother had 'come up with this interesting thing'.

---

* Nowadays he's more relaxed about publicity. 'I'd realised that the engine could become something, and it was up to me to help make it become something' – and if that includes indulging TV crews and authors, he's now happy, in principle, to do so.

Berger took a look and agreed (the car helped) – although there was still a good deal of head-scratching as to where the engine might find a market. Despite Peter's advances in efficiency, the Dearman Engine was never going to compete with petrol or diesel for sheer power. Nonetheless, Berger put up the £30,000 Peter needed for a patent and introduced him to ex-warzone photo-journalist turned eco-entrepreneur Toby Peters, one of the two 'nutters' who would later go on to wow the cynical Tim Fox. With Toby on board, the three men began to look for a solid, commercial application for Peter's invention.

When they worked out what it was, it was huge.

The United Nations Food and Agriculture Organisation estimates that one in eight people go to bed hungry. And the world's population is still growing, though thankfully at nothing like the rate it was when Norman Borlaug kicked off the 'Green Revolution'. Nonetheless, The UN's 'mid-range projection' for global population estimates the human race will reach nearly 11 billion people by the end of the century and then level off (worldwide fertility rates have been dropping for over fifty years and on current trends will hit 2.1 – replacement rate – sometime towards the end of century).* But that's an extra 4 billion mouths to feed by the end of the century, 80% of those arriving by 2050. Even commentators who aren't versed in the sustainability nightmare that bedevils the Green Revolution are calling 'Armageddon!' If one in eight of us go to bed hungry now

---

* The reasons for this are many, interrelated and hotly debated, but a defensible top-level summary is that, as nations become more prosperous, their populations urbanise, which brings with it female emancipation, and the combined effects have a downward influence on birth rates.

(that's about 900 million people), how can we possibly hope to cope with the additional billions soon to come?

It's a problem that Tim Fox at the Institution of Mechanical Engineers has thought about a lot. Over lunch we talked through the headlines of a report he'd authored on world food supply. Released in January 2013, its findings dominated the news headlines that week because it brought into stark focus another huge problem with our global food system. Titled *Global Food: Waste Not Want Not*, it estimated that between 30% and 50% of the food we produce each year *never reaches a human stomach*. That's a staggering 1.2 to 2 billion tonnes of food we grow, but never eat. Every year.* To put that in context, if a billion tonnes of food was expressed in uneaten bananas laid end-to-end, the line would stretch past Saturn. The amount of waste is colossal.

In sub-Saharan Africa, for instance, averaged per person, 167kg of harvested food never gets eaten. 7kg of that is consumer waste, but the other 96% is lost between farm and customer during the harvesting, storing, processing and distribution of that food. This 'post-harvest loss', as it's called, accounts for approximately *a third* of all that's grown in the area.† When you consider that sub-Saharan Africa has the largest percentage of hungry people in the world, at just under a quarter of the population, you begin to understand what a senseless tragedy this is. Hunger kills more people than AIDS, malaria and tuberculosis combined, and whilst it's important to note that not all hunger is down to waste (poverty, climate change, war and volatile food markets all

---

* A report published by the UN's High Level Panel of Experts on Food Security and Nutrition puts the figure at 33% of all food – or, by their estimate, 1.3 billion tonnes.

† The figures aren't much better elsewhere. Farms, wholesalers, distributors and retailers account for 93% of waste in South and Southeast Asia. In Latin America nine out of every ten tonnes of food wasted is lost *before* it reaches a dinner plate, not *afterwards*.

also play a role), if we eliminated post-harvest losses, we'd make a major dent in the problem and have a much better chance of feeding our growing population.

It's not only the lost food itself that needs to be considered. The associated waste of production resources – all the water, energy and human toil that went into producing it is eye-watering. 'About five hundred million cubic metres of water, which is about an eighth of what we consume in the world, is lost on crops,' says Tim.

I find myself thinking back to my time in Jharkhand, learning of the rapid retreat of the water table. It's bad enough that we've created a farming system that's decimating our groundwater reserve, even more upsetting to realise that we're literally throwing away a huge proportion of the water we so assiduously seek to extract. (For a species that is 60% H_2O, we're dangerously and depressingly cavalier with the stuff.) And besides the water squandered, food wastage also accounts for 3.3 billion tonnes of greenhouse gases, 1.4 billion hectares of land use and an estimated loss to the world economy of $750 billion *every year*.

It turns out that a huge proportion of the food waste suffered in 'developing' economies is down to poor refrigeration. Without it, fresh fish lasts for a day, milk a few hours, meat no more than two days, and most fruit and vegetables less than a week. The result is that developing nations' economies are put at a huge disadvantage in terms of nutrition, healthcare (medicines, blood and soft tissues all need refrigeration, too) and economic development. This problem is exacerbated by the fact that these countries are generally the hotter ones – and, when it comes to food and medicine, the hotter it gets, the quicker it rots, wilts and spoils.

This is why the 'developed' world spends a veritable fortune on refrigeration, roughly a $100 billion a year, sustaining an 'artificial cryosphere' – networks of refrigerated warehousing and transportation that get food from farm to fork with minimal loss. To give you an idea of how big the cold industry is, 16% of *all* electricity consumed in the UK goes on cooling. Investment in cold is one of the pillars upon which prosperity rests, a key foundation of a modern society. So nations across Africa, Asia, the Middle East and South America are understandably keen to develop 'cold chains' of the standard enjoyed (but rarely noticed) by the citizens of richer nations. It's something I talked about with Yezdi as we'd walked among the music-listening cows in KGVK's dairy.

'There's no real sustainable cold chain here, or in India generally,' he'd told me, which means post-harvest loss is a big problem there. The country only has 7,000 refrigerated trucks for transporting perishable produce, which when you're trying to feed a nation of 1.2 billion people is a major problem.

It was Toby Peters, with his experience as a journalist across Africa, who had the next brainwave. The almost forgotten characteristic of Peter's invention – that it ran very cold (minus 195°C) – meant that, with a little engineering nous and utilising already well-established technologies, you could create an engine that, while producing power, also provided a refrigeration service for free. Toby realised that Peter had unwittingly come up with a new technology for refrigeration – and one far greener than the fossil-fuel-powered methods we use today.

'After we realised we could use it for cooling, things just snowballed,' Peter told me (which I think may have been a pun).

I'm at the headquarters of the Institution of Mechanical Engineers, the same building where Tim Fox first met Toby and Gareth. As he entered that meeting he could not have imagined that a few years later he'd be pulling together the summit I've come to attend – an international gathering to coincide with the publication of the institution's latest opus: *A Tank of Cold: Cleantech Leapfrog to a More Food Secure World*, a report on the very technologies he'd been so reluctant even to hear about.

Peter Dearman is a notable absentee (he's not one for crowds) but just about everyone else interested in a new cold revolution is here. The list of attendees includes post-harvest loss experts, refrigeration specialists from large supermarket chains, agricultural attachés, various government 'development authorities', directors from manufacturing firms, management consultants, corporate financial advisors, representatives from industrial gas providers, even Chinese environmentalists, and, as you'd expect, Toby Peters; now CEO of the seventy-strong Dearman Engine Company. The buzz in the air is palpable.

Of the presentations given in the Institute's grand lecture hall one in particular catches me. It's delivered by Lisa Kitinoja, founder of the Post-Harvest Education Foundation, based in Oregon. She's a world expert on post-harvest loss and wants to talk about Tanzania, a country where the temperature rarely drops below 20°C, yet which has no cold chain to speak of, despite an economy dominated by agriculture. Post-harvest losses are 40% or higher, she says – so you can see why she's excited about Peter and co. In my hand is Lisa's thirty-page analysis of what Dearman's technology could do for the developing world.

It makes for hopeful reading. She imagines farms where liquid air systems provide power and cold at the same time, running, for example, a refrigerated conveyor belt for fruit processing. In one example she estimates a Dearman system could raise the wholesale price for a tonne of mangoes from $100 per tonne to an astounding $3,000. Mangoes that would have been lost to the punishing heat could now be chunked and frozen (chunked mango is a premium product), generating revenue throughout the year. The ability to minimise waste delivers a massive economic dividend.

And, while today that cooling and energy could possibly be provided more cheaply using a diesel generator, Lisa, in common with the assembled engineers and policymakers here, is excited by the idea of a new 'cold economy' that, at scale, will challenge the cost of fossil-fuel-powered alternatives and come with three key advantages that are hard to compete with.

The first is that liquid air is a fuel source that every nation can make for itself. Tanzania has no domestic oil industry to speak of, so any attempt to create a nationwide cold chain using existing technologies would be dependent on fuel imports as well as victim to the vagaries of fluctuating oil prices.* Of course, you need energy to compress the air, and that energy could come from fossil fuels but, as I find out later on my travels, in the developing world that power is increasingly going to come from renewable sources. Meanwhile developed nations are currently throwing away vast quantities of liquid nitrogen (which the Dearman engine runs equally well on) because there's nearly four times as much nitrogen as oxygen in the air, but currently much more commercial demand for the latter. Now, thanks to the Dearman Engine that wasted liquid nitrogen has become a near-free fuel going begging. Uber-geeks might like to know that it takes about 439kWh of electricity to create a tonne of

---

* The same is true for Benin, Botswana, Burkina Faso, Burundi, Central African Republic, Djibouti, Eritrea, Ethiopia, The Gambia, Ghana, Guinea, Guinea-Bissau, Kenya, Lesotho, Liberia, Madagascar, Mauritania, Morocco, Mozambique, Namibia, Niger, Rwanda, Senegal, Sierra Leone, Somalia, Swaziland, Togo, Uganda, Western Sahara, Zambia and Zimbabwe.

liquid air. In a Dearman Engine this will yield you 120kWh of mechanical energy and 100kWh of cooling energy – or an efficiency of 50%. That compares very favourably to the modern gasoline engine which delivers 30% efficiency, if you're lucky. (Yes, 70% of the energy you pay to put in your tank is lost, mostly dissipated as friction and waste heat).

The second benefit is that liquid air refrigeration could significantly reduce air pollution. Refrigerators across the cold chain are often powered by diesel engines – engines that typically remain under the radar when it comes to clean air legislation. This means, for instance, that while the engine driving a food truck is regulated by law, the secondary diesel engine powering the refrigeration unit, and travelling on exactly the same journey, isn't. An analysis by the sustainable energy consultancy E4tech concluded that diesel-powered refrigeration units can emit six times as much nitrogen oxide and a cough-inducing 29 times as much 'particulate matter' (a catch-all term for particle pollutants) than a regulated lorry engine does. There are 4 million refrigerated trucks worldwide today, a figure set to grow to between 10 and 17 million trucks by 2025, growth which threatens to exacerbate a pollution problem that's already out of control in many of our cities. Dearman engines, by comparison, emit nothing but clean air. Nick Owen, one of the co-authors of the research for E4tech, was so convinced by the technology he quit his job to become Chief Technology Officer for the Dearman Energy Company.

In the coffee break, Tim outlines the third advantage liquid-air solutions have over the existing refrigeration technologies. It's the one he regards as the big kahuna, a benefit that dwarfs (though is related to) those of not being dependent on oil imports or contributing to air pollution.

'The global South is going to pay heavily if we don't do something about climate change,' he says. 'So how does it make sense to put in a system to cool things down that, by its very operation, will warm things up?! Existing cold chain technology chills our food by heating our planet. Why would you do that, when we now have a better alternative?'

I'm paying a visit to the Motor Industry Research Association (MIRA) test facility on the site of an old RAF airfield outside Nuneaton. It's *the* proving ground in the UK for new vehicles, with miles of simulated roadways. If you're a *Top Gear* fan you may have seen the show's presenters try out some of the motor industry's latest offerings here. Not that I'm allowed to film anything. Without a special permit you're banned from using photographic equipment. My smartphone camera lens is taped over with a tamper-evident sticker on my arrival. MIRA's clients are understandably nervous about the press or the competition seeing their works-in-progress. Not only are cameras banned, but vehicles under test often sport bizarre black and white swirly paint jobs, designed to outfox the auto-focus systems of telephoto lenses. Indeed a supercar-meets-zebra hybrid passes by the hangar doors of the Dearman Engine Company's temporary home here, where I've come to see if this new technology can work outside the lab and on a real vehicle: the world's first food truck refrigerated using one of Peter's engines.

The truck itself is nothing to write home about. It's an old specimen, although it does have a new number plate that reads 'Cool-E' – a reference to the consortium, backed by funding from the UK government, that includes MIRA itself, the Dearman Engine Company, Loughborough University and Air Products, a huge multinational that sells gases and chemicals to industry. It seems plenty of people are already convinced about the potential. On the train up I'd read a report by the Centre for Low Carbon Futures – a multi-university research consortium looking at commercial solutions to environmental problems. They'd concluded that Dearman's technology could soon deliver £1 billion in annual revenues and 22,000 jobs to the UK.

I'm introduced to two fresh-faced engineers – (another) Tim and Scott – who are busy putting the vehicle through its paces. Tim tells me he's long been trying to make the move into green technologies. 'Before, I was working for *Ford* but it felt like a contradiction. I was trying to make diesel engines more efficient but at the end of the day I was still making diesel engines, which are not good for the environment.' Disillusioned, he'd decided to quit his job and go travelling, when the Dearman Engine Company intervened. 'I saw this advert for a job to work on a zero emission engine – and it was like: "hang on!"'

Slung beneath the truck is a descendant of one of the two Dearman engines I'd seen in that basement laboratory at Imperial College. I ask Tim and Scott to talk me through what I'm looking at. 'So the truck still uses a diesel motor up front,' explains Tim, 'but we've swapped out the second engine, the one that powers the refrigeration unit, for a Dearman Engine.' 'Just *running* the Dearman engine creates cold,' adds Scott, 'and that's fed into the back of the truck, providing about two-thirds of the refrigeration needed.' 'The rest of the cooling is generated using a traditional refrigerator,' explains Tim. 'But the *power* for that refrigerator is coming from the Dearman engine, too.' Scott again: 'Also bear in mind that because most of the cooling comes for free, the refrigeration unit is running far less than it would on a normal system so it'll last longer as well as being powered by completely clean energy.'

They're both smiling despite, it turns out, working long hours. 'The workload is quite stressful, and our timelines are quite tight,' says Tim, 'but I feel this is a time I'll look back on and think, "Yeah, that was really good. The opportunity we've got ..."' Scott chips in. 'It's hopefully world-changing. It's something that hasn't been done before and that's pretty cool.'

Tim checks a readout which tells him the truck's chilled compartment is sitting at minus 6°C, his breath condensing in the cold. On other test runs the truck has happily maintained a temperature of minus 20°C. When the truck finishes its time here it will be handed over to the supermarket chain Sainsbury's for an initial commercial deployment.

When Toby Peters first came to see Tim Fox, the latter had almost dismissed his visitor out of hand, but with an annual world market for refrigeration equipment already around the $35 billion mark (and set to expand rapidly), there's a very real possibility that Toby might end up being the Henry Ford of Cold. But the story doesn't end there. Because it turns out that Peter Dearman's antifreeze brainwave has *another*, potentially world-changing application. The man in the shed might have helped change the world *twice*.

6 INSTANT POWER

'Energy and persistence conquer all things.'
– BENJAMIN FRANKLIN, POLYMATH

You can often find Professor Yulong Ding at Birmingham University's School of Chemical Engineering, the latest in a long line of academic homes for one of the world's most published scientists. Professor Ding is right up there – a research celebrity of sorts – and the very picture of a modern academic. Thin-rimmed spectacles, a mop of black hair and the strong impression that even as you're speaking to him some deep processing is going on in another part of his brain. A quietly spoken Chinese immigrant, he is a long way from his rural childhood home in Jiangsu Province, where, despite poor schooling, he became a star student, driven not by parental aspiration or expectation (his mother still does not read or write) but an all-consuming curiosity. Physics, chemistry and mathematics enthralled him.

'I just love mathematics!' he tells me and I believe him.

Top marks at school were his passport to become 'one of the two or three per cent of people at the time selected to get into university', where, as the 1980s dawned, he found himself studying the physics and mathematics of energy at Beijing's prestigious University of Science and Technology, mostly in the service of the Chinese steel industry. He admits to working six and half days a week solidly for years.

'I didn't really go out much. I'm pretty sure I was top of the class.' I'm pretty sure he's right – the office we're meeting in is festooned with awards. Professor Ding is an incredibly smart cookie and, to be honest, he knows it. It's not arrogance; rather a certain sure-footedness in the way he thinks and talks that I've witnessed when meeting other super-brains.

Following his degree, he was recruited to investigate ways to clean up coal combustion. This wasn't because China was overly worried about air pollution or climate change at the time (although they are now), but because impurities could clog up machinery, making it less efficient and prone to malfunction. He did, despite his work ethic, find time for love, meeting his wife (an intensive care nurse specialising in heart conditions) at one of the few parties his colleagues dragged him to. A son followed, a key catalyst for Yulong to leave China – not for ideological or political reasons (he retains a strong affection for his homeland), but logistical ones.

Despite working for a further six years studying the thermodynamics of metals (and receiving many accolades for doing so), Yulong and his family were still crammed into a tiny one-bedroom apartment (generously vacated by two colleagues who were supposed to share it with them), with no prospect of anything more roomy on the horizon. 'Proper apartments were allocated according to the years of service, not your contribution, and there was a long queue ahead of us, so I applied for a visa to work abroad.'

Thanks to a UK Overseas Research Student Scholarship, coupled with money to cover living expenses, Beijing's loss was Birmingham University's gain. For want of a bedroom, the UK acquired one of the smartest minds on the planet.

Arriving in Britain in the early 1990s, Yulong found life 'initially very hard'. His written English was good but he struggled with the spoken language and suffered occasional episodes of racist bullying (both in work and outside), although he says on the whole the people of Birmingham were 'very friendly'. Two decades later he's much happier. His English is confident and conversational and he's integrated with British society enough to admit that he's a fan of cheese, local beer and even (surely this must be heretical back home?) English Breakfast Tea. The intervening period has seen his career continue its impressive trajectory. Via stints at one of my previous ports of call, Imperial College (where he did some of the first work on capturing CO_2 from industrial processes), and Leeds University (heading up a department looking into nano-thermodynamics, amongst other things), he's now back in Birmingham, where he's spent a large part of the last decade investigating the possibilities of liquid air – 'the big story' as he calls it.

'So, one day in 2005 I'm sat in my office and Toby Peters comes in and starts talking about a man called Peter Dearman …' Yulong's reputation in thermodynamics made him the go-to guy to validate the science behind the Dearman engine.

'Were you initially sceptical?' I ask.

'No, I wasn't. I met with Peter, and looked at what he was doing and thought it was perfectly possible. We get on very well.'

Yulong wants to show me the result of the conversations that followed, the second large-scale application of Dearman's trick with the antifreeze. We leave the office, passing a number of laboratories in which the odd researcher is asleep at their desks (they obviously work them hard in Birmingham) and through some double doors taking us outside, where I find myself looking at four shipping containers and a couple of huge gas storage tanks linked by a maze of ducting. The whole affair is the size of a large detached house.

'This is it?' I ask.

'Yes,' says Yulong. 'This took about ten years of my life, and I'm very proud of it.'

If you fancy a bracing walk and happen to be in north-west Wales, I recommend taking a steep hike up Elidir Fawr. The best part of a kilometre high, it may be one of the lesser known mountains in Snowdonia, but to me it's the most interesting – because as well as being a mountain, it is also one of the world's largest batteries.

Like all energy grids, the UK's National Grid maintains a host of energy storage facilities that it can call into action at times of high demand or fuel shortages. At the moment 99.92% of that storage takes the form of massive stockpiles of gas and coal. Only a measly 0.08% is rechargeable storage, some of which you can find at the top of Elidir Fawr in the form of Marchlyn Mawr reservoir. If you're on the mountain at the right time you might see the strange phenomenon of it filling up, even if it's not raining. What's happening is excess power on the grid is being used to pump a huge volume of water into the reservoir from Lake Llyn Peris half a kilometre below. If the UK suddenly needs some extra power,* the water can be released, passing through Europe's largest man-made cavern (known as 'the concert hall' and big enough to house St. Paul's Cathedral) that contains six enormous turbines.

Up to 7 million cubic metres of water gush through these gargantuan machines generating an impressive 1,650 Megawatts of power – roughly three times as much as your average coal-powered power plant. If Marchlyn Mawr is full, this monstrous

---

* For example, during commercial breaks in popular TV programmes, when the nation decides to make a cup of tea rather than be marketed at, something referred to in the energy trade as a 'TV pickup'.

rocky battery can discharge power for five hours straight. Time from a standing start to peak output? About 16 seconds – a marked contrast with conventional power stations, which can take 12 hours or more to power up from cold.

Elidir Fawr (also known as Dinorwig Power Station, or 'Electric Mountain') took a decade to build and cost £425 million, which actually turns out to be pretty good value for money. Because of its huge capacity and long operating life, per megawatt, it's one of the cheapest forms of rechargeable utility-grade energy storage available. However, pumped hydro systems are not the sort of thing you can build quickly or without a good deal of pesky admin. Finding a mountain, hollowing it out, building a power station that can operate underwater inside and paying for the whole thing is a major exercise, which gives the world a problem. Because we desperately need a lot more batteries, and we need them soon.

In the battle against climate change, nations need to decarbonise their electricity systems, bringing more renewables online. But there's a sticking point, particularly for countries in the (less sunny) northern hemisphere. Solar panels only transform sunlight into electricity when the sun is shining and wind turbines when it's windy – something referred to as the 'intermittency problem' and a popular stick the fossil fuel lobby uses to bash the renewables sector with. An energy system based on renewables is too uneven, they argue, generating energy at the whim of nature, not man – too much energy when it's sunny or windy, none at night or when it's still.

The solution to intermittency is grid-connected batteries that can store the excess and smooth out supply, discharging their bounty when needed. A decarbonised energy system simply won't work without good storage, and the more nations look to attach greener energy to their grids the bigger the challenge gets.

It's also a problem for countries without national grids, like Tanzania, where only 15% of the population have access to electricity (in rural areas that figure drops to 2%). With an annual average personal income of $250, there are few prospects that 'the financial resources will become available for Tanzania Electricity Supply Company to undertake electrification of even 20% of the rural households in the foreseeable future'. It's a story repeated across sub-Saharan Africa, where the majority of the 1.3 billion people without access to electricity live – one reason why community-owned solar power is burgeoning across the continent. As solar panel prices continue to fall, installations become financially viable for households and communities living in some of the sunniest countries on Earth.

Just as rural Africa bypassed fixed networks when it came to telephony, so it could with electricity, but it will *only* work well if we solve the intermittency problem. Those communities need the electricity through the night, which means they need batteries. Hydro-pumped storage isn't an option and electro-chemical batteries are simply too expensive.

The good news is that batteries are what futurists call 'an exponential technology', that is, one that increases its capacity whilst reducing in price at regular intervals (computer processing power being the most famous example of this phenomenon). An analysis by Duke University of battery price trends per kilowatt-hour stored led futurist Ramez Naam to conclude that battery technology will be 'cheap enough to displace most fossil fuel use for electricity' by 2025. This chimes with an analysis by UBS, the world's second largest bank, who estimate that electrochemical batteries will drop fourfold in price in a decade. Both analyses are based on the reasonable assumption that increasing demand will continue to drive down costs. But that increasing demand brings with it environmental worries. A life-cycle assessment of lithium-ion batteries (expected to form the bulk of the utility energy storage market) by the US Environmental Protection Agency highlighted very real concerns of resource depletion (especially of rare metals), 'ecological toxicity' (driven by metal ore extraction and processing) and several human health impacts both to the general public and workers in the battery industry – notably worries about exposure to carcinogens. Batteries may be essential, but they're dependent on a host of toxic components.

Unless, of course, you make them out of air.

'**We were thinking about how you might use a Dearman engine** to cool food on a long sea journey,' says Yulong as we walk around his obsession of the last ten years. 'In that situation you'd need to take a big store of liquid air with you.'

And then it hit them: Peter had made it economic to use liquid air as a storage medium for energy. Thanks to his trick with the antifreeze you can now get far more energy back out of a tank of air than before, creating a widely deployable battery technology at the same price point as Electric Mountain but without any of the environmental worries that beset electrochemical solutions.

'So, I'm looking at a massive rechargeable battery?' I say.

'Yes,' replies Yulong. 'A battery you can make out of standard engineering components, put anywhere in the world, that never loses capacity and stores energy in a completely non-toxic way.'

This first-of-its-kind battery is a recent arrival in Birmingham, having spent the last three years on an industrial estate in Slough, where it was put through its paces as part of the National Grid's 'Short Term Operating Reserve' (STOR) contract – one of the first major projects for the newly formed Highview Power, a company run by Gareth Brett, the other 'nutter' that came to see Tim Fox all those months ago. The university will soon put the new arrival to use servicing its own energy needs.

Much of the talk at Tim's 'Clean and Cool' summit had been about a new 'cold economy' – a suite of integrated applications including energy storage, refrigeration, air-conditioning, even small air-powered zero-emission vehicles – all of which could work stand-alone or slot into our existing infrastructures. People like Lisa Kitinoja are excited by the idea of solar-powered air liquefiers located in or near rural towns, where citizens can use stored liquid air either to chill food or generate electricity. Farmers from nearby villages can collect liquid air to run their own smaller operations. Long-haul deliveries are achieved using a standard diesel truck, but with a Dearman-powered refrigerator keeping the produce fresh.

In larger towns, hotels and offices can use liquid air technology for air-conditioning, the impact of which shouldn't be underestimated. Lee Kuan Yew, the first prime minister of Singapore, transformed his homeland from a poor, tiny third-world island port with no natural resources to the modern city-state acknowledged as a world leader in free enterprise (ranked second in the world by the 2015 Index of Economic Freedom) and science and enjoying per capita GDP figures higher than the USA or UK – and he did all this within fifty years. Interviewed in 2009, he was asked what he thought the foundations of Singapore's success had been. His answer? Multicultural tolerance ('we are a conglomeration of people who were thrown together by the British') and air-conditioning:

> '*Air-conditioning was a most important invention for us,*
> *perhaps one of the signal inventions of history. It changed the*

*nature of civilization by making development possible in the
tropics. Without air-conditioning you can work only in the
cool early-morning hours or at dusk. The first thing I did upon
becoming prime minister was to install air-conditioners in
buildings where the civil service worked. This was key to public
efficiency.'*

Singapore is a nation that exists because it can keep itself cool.

At the Liquid Air summit I'd attended at the Institute of
Mechanical Engineers there were rumours of a major tuk-
tuk manufacturer talking to the Dearman Engine Company.
Smaller vehicles could be entirely powered by a Dearman engine
– a zero-emission form of transport that could house a chiller
compartment in the back for in-city deliveries, while providing
an air-conditioned cab, and reducing urban air pollution into
the bargain. Indeed, Tim Fox has a personal liquid-air ambition
to compete in India's biannual Rickshaw Run in the world's first
liquid air-powered tuk-tuk. The race is described by its organisers
as a '3,500km pan-Indian adventure in a seven-horsepower
glorified lawnmower' and 'easily the least sensible thing to do
with two weeks'. 'It was my daughter Shannon's idea. She's going
to do it with me,' he tells me during a phone catch-up in his
new role as 'International Ambassador' for, you guessed it, the
Dearman Engine Company.

'So you're one of the nutters now?' I ask. 'How is it?'

He laughs. 'In a word: busy. We'll be testing the first commercial
trucks and putting together projects in Spain, the Netherlands
and Malaysia. Singapore is keen to refit its air-conditioning to be
more climate-friendly so we're in conversations there, too. We're
also in discussions in Hong Kong about air-conditioning their

buses. California and New York State are competing to clean up their act and we've been out talking to the governments of both.'

I can hear the excitement in his voice and it's catching. When Dr Tim Fox, one of the UK's most senior and respected mechanical engineers, quits a job working for his profession's most prestigious institute to bet his career on a new technology it means something.

'You seem happy,' I say.

'I am! Look, if we get it right, this technology is a game-changer,' he says urgently. 'I've been an engineer all my life and I know this stuff works and it can change things for the better for millions of people.'

It's another piece in the jigsaw puzzle of how we might reboot our systems that's beginning to emerge on my travels. I've investigated challenges to the status quo in healthcare, agriculture and now, with Yulong and Peter's air battery, one part of our energy system. But there's a lot more to re-imagining energy than changing our storage technologies. The world needs a wholesale rethink of how the power we use is generated and owned if we're to have any chance of making the future more sustainable, equitable, humane and just. It's why I'm heading to Austria, to hear the story of a town back from the brink, and to meet one of the most invigorating people on the planet.

7 EDISON'S REVENGE

'Some people want it to happen, some wish it would happen, others make it happen.'
– MICHAEL JORDAN, SPORTSMAN

Meeting Peter Vadasz is one of the best things I've ever done. The ex-mayor of Güssing, a picturesque town of 4,000 souls in the Austrian province of Burgenland, knows what it takes to live a full, happy and meaningful life. Joy, enthusiasm and gratitude radiate from him so freely that after a few moments in his company you feel happy to be alive and ready to take on the world. His company is like a lightning rod of optimism. As we talk, everyone who passes by raises a hand or shouts 'hello'. He waves back, exchanging a few words, asking after people's children or how a local business is doing. It's obvious he's respected and well liked. But it wasn't always this way. In fact, Peter recalls a time when plenty of people around here treated him with contempt.

Born out of wedlock in 1944, he was raised by his grandmother in a single room with no electricity. 'A lot of people looked down on us because we were poor. But that time of extreme poverty formed me,' he says. 'My grandmother taught me you have two choices. Either you say, "OK, wait until I'm up there, I'll show *you*," or you say, "*Never* do this to anybody else," And the latter is the better because, believe me, the worst thing in life is hatred. If you start feeling hatred towards other people, it will kill you.'

Poverty was commonplace in Peter's youth. Residents still talk in dim tones of 'fifty years alongside the Iron Curtain', remembering only too well times when the town was dying a slow and seemingly inevitable death, one of the latest indignities visited on an area that's often found itself inadvertently at the crossroads of conflict. On the dividing line between East and West Europe, the district and its bijou capital has been the victim of hastily redrawn borders on more than one occasion, giving Güssing and the surrounding province, as the receptionist at my hotel put it, 'a bit *too much* history'. We're seated in the shadow of one bit of that history, meeting at a bar-restaurant in the grounds of twelfth-century Burg Güssing, the castle atop an extinct volcano, around which the town is built. It's a gloriously sunny afternoon, and yet the weather is soon outshone by Peter's easy laughter and bonhomie.

Peter tells me how he became a teacher of history, geography, and English (explaining how he speaks my native tongue so fluently) and actually taught alongside his successor, the current mayor. During this period he also met his wife, who, he claims with typical playfulness, 'seduced me, about 10 kilometres from here'. He points. 'In that direction!'

Why, I ask him, did he get into politics?

'I wanted to change something,' he tells me. 'If you see the poverty, if you see the migration out …'

'It wasn't a career choice?'

'Forget career!' he snorts. 'If you go into politics just for career, you will never achieve anything really lasting.'

'Do you miss it?'

'No! I tell you, retirement is the best thing that's ever happened to me! What a wonderful feeling!'

It's a striking statement for two reasons. The first is that he doesn't look nearly old enough to be taking his pension. His skin is that of a man in his late forties, his eyes sparkle like a curious infant and he smiles about once every ten seconds. The only giveaway is the thick silver hair, which along with his slightly pointed features gives Peter something of an impish quality, the whole package reminiscent of a playful wizard. When I remark on his youthfulness he asks, 'Do I have to buy the beers now? You should go into politics!'

The second reason is that later on in our conversation he will tell me that his time as mayor was also 'the best thing that ever happened to me' and that meeting a man called Herman (whose ground-breaking work I'll marvel at while I'm here) was, quite definitely, 'the best thing that ever happened to me'. This, I learn, is Peter Vadasz in a nutshell: he finds the best in every situation and it's an attitude that's infectious. I'm beginning to think that sitting drinking beers with this cheery former mayor is certainly one of the best things that's ever happened to *me*. He'll make you believe anything is possible, which given that under his leadership the town challenged the powerful forces of the energy industry and came out smiling, doesn't seem such a mad thought.

Even the Terminator, Arnold Schwarzenegger, was impressed. Visiting in 2012, he concluded, 'The whole world should become Güssing.' That wasn't just hyperbole: Schwarzenegger (in his role as governor of California, not a time-shifting robot from the future), along with many others, believes Güssing may have found a model of sustainable prosperity that redefines who pulls the strings in the economy. It's not a bad endorsement for a place that two decades earlier was on its knees. Schwarzenegger is perhaps the most famous of the (up to 20,000) visitors who arrive every year from every corner of the globe to, as Peter puts it, 'convince themselves that what they have read or heard is actually true'.

'When I was elected in 1992, we were facing a lot of problems. Young people left the area because we had no jobs and no future for them.' Those who did secure what scarce employment was on offer suffered the lowest income rates in Austria. Pretty much everybody else was either unemployed or commuting to Vienna, 160 kilometres north. The local economy was in permanent freefall. But today the town is reborn, bolstered by fifty new companies, bringing over a thousand new jobs, drawn to the area by an idea that was born fifty metres from where we are sitting, thanks to another life lesson from Peter's grandmother.

'She always told me, "Look Peter, I have you. You have me. Don't concentrate on what you don't have. Look at what you do."' He points to a spot overlooking a set of battlements where, early in his mayorship, he stood with the other key architect of the town's rejuvenation as they asked themselves what resources the area *did* have at its disposal. I take a look for myself. The answer is obvious. From Burg Güssing you can't avoid the fact that the local district has one thing in abundance.

Trees. Lots and lots of trees.

At the root of the town's problems was an issue so embedded in day-to-day life that people rarely questioned it. And what was once true in Güssing remains true for the rest of us.

'Every one of us pays for energy,' says Peter. 'But the question is, where does the energy come *from* and where does the money *go*? Does it go to Saudi Arabia, does it go to Russia, with love?'

The year before Vadasz became mayor, the citizens and businesses of Güssing spent over €6 million on energy, mostly on oil (for heating) and utility-supplied electricity. All that money was flowing out of the local economy, as it does pretty much the world over.

If you ask yourself where the money you spend on energy goes, it's unlikely to be anywhere nearby. If you live in Hong Kong, for instance, 99.5% of your energy is imported. Austria itself imports over 60% of its power. In the UK, net imports of energy stand at about 44%, a figure that's set to rise. My nation's extra coal comes from Russia, Colombia and the USA, our extra oil and gas largely from Norway, our extra electricity from France and The Netherlands.

Even if energy is generated locally, the profits from doing so are spent elsewhere. In the UK, the money we pay to heat and power our homes and businesses mostly goes to the employees and shareholders of one of the 'big six' energy companies (three of which are foreign-owned). Between them they hold a 92% market share.

The USA can at least claim to keep the majority of the money its citizens spend on energy within the domestic economy. A home-grown shale oil and shale gas boom, along with improved energy efficiency, means US energy imports are plummeting – from nearly a third of all energy in 2005 to about a sixth today and, if the US Energy Information Administration's predictions pan out, to only 4% by 2040. As is the case elsewhere, the spoils go to employees and shareholders – and you might not think that's a bad thing. After all, if those companies have gone to all the trouble of providing the energy that powers your devices and cooks your evening meal, they should be rewarded, shouldn't they? Without them nothing happens.

The problem for Güssing was that the money flowing out to pay for energy was further crippling an already stalled economy. Doctor's surgeries, schools, shops and pubs were becoming impossible to sustain. The town simply couldn't afford to lose that €6 million each year. The answer? Güssing was going to generate its own energy. Not in a piecemeal fashion, a solar park here, or a wind turbine there. No, the mayor's office set the goal of taking the entire town over to 100% self-generated energy.

This was seen as an heretical, foolish ambition, especially back in 1992, but it's Güssing's early start down the energy-independence road that makes it so fascinating to all those international visitors. Ahead of my visit, Roswitha Gruber at Güssing's European Centre for Renewable Energy sent me a brace of documents covering the town's energy projects over the last twenty-five years. Whilst the tone was one of justified pride about what the town has achieved, there was no shying away from political difficulties, the questionable economics of particular projects, or the need for constant improvement and experimentation. If you want the straight dope about locally generated energy, there probably isn't anywhere better to visit. The hard truths about what it takes to change energy, to take on the existing system, and win, are to be found in this quiet town nestled by the Hungarian border.

Güssing's success is as much the story of two men's tenacity as it is of any technology or business model. The first of them is sitting opposite me, a man born into poverty who ended his career as mayor. The second will enter our story shortly, a middle-class kid who became a local hero – two boys from different sides of the tracks, but with one crucial attribute in common.

'We were both a little crazy,' says Peter, chuckling.

How did we get the energy system we have?

Some history. In 1882, Thomas Edison opened the first central power plant in America to supply electricity to a new kind of lightbulb – one that used thin carbon filaments, which glowed hot as they resisted the electricity flowing through them.* Until then, electric light usually came in the form of carbon arc lamps that worked by creating 'mini-lightning' between two charged carbon electrodes. These unwieldy 'lightning' lamps illuminated streets, public buildings and factories. They had two main disadvantages, however. They were expensive to maintain – the electrodes slowly vaporised as they worked and had to be regularly replaced – and they were often deadly. The extreme heat they emitted (the reaction inside a carbon arc bulb generates roughly 3,600°C) and the sparks they were prone to produce made them a fire hazard. (The 1903 Chicago Iroquois Theatre Fire – the deadliest fire of its kind in US history – was started by an arc light igniting a muslin curtain. The tragedy claimed over 600 lives, the bodies piled in the street.)

The high voltages required by the lamps were another consideration, meaning the wires supplying them had to be placed underground or safely distant from other overhead cables. Except they often weren't. In New York, for instance, crooked board members at the Board of Electrical Control handed out lucrative contracts to relatives and friends who suspended the wires right next to those used by the telegraph, resulting in a number of grisly and public electrocutions. Perhaps the most

---

*   Edison is often attributed with the invention of the filament lightbulb, but this simply isn't true. Almost simultaneously Joseph Swan in the UK created much the same device and before both of them at least twenty-two other inventors had created bulbs based on similar principles. What can be said is that Edison's version was arguably the best of the bunch, But his success (and therefore his association in the public consciousness with the lightbulb) was as much down to his skill as an entrepreneur and system innovator as an inventor – thinking not only about the construction of the bulb but also about how a whole network of lighting would work, including generating the power necessary to illuminate his creations, distributing it and working out how to get paid for doing so.

gruesome of these occurred in 1889, just before Mayor Hugh Grant (yes, really), in response to such tragedies, plunged much of the city into darkness after ordering the main suppliers of municipal lighting to switch off their services. The *New York Times* reported how maintenance engineer John Feeks, reaching through a tangle of wires, grabbed an illegally placed live cable and, 12 metres above ground, became a ball of fire. 'Blue flames issued from his mouth and nostrils and sparks flew about his feet. Then blood began to drop down from the body' as he 'hung in the fatal burning embrace of the wires.'

All this was grist to Edison's mill. His incandescent filament bulbs could be powered at much lower (and therefore, he argued, safer) voltages. As the death toll rose, Edison was only too happy to demonstrate the superiority of his Pearl Street Station, which had, at the time of Mayor Grant's blackout, been safely powering over 10,000 lights serving hundreds of local customers for the last seven years (customers who included the *New York Times*).

But Edison had a problem. Nearly all materials resist electricity, a phenomenon Edison had used to his advantage in his bulbs, whose carbon filament manifested their resistance as the light he sold to his customers. But resistance elsewhere in a circuit (even the comparatively low resistance offered by copper wire, for instance) will, over a large enough distance, take its toll on the current, so called 'transmission losses'. In practical terms this meant that to be one of Edison's customers, you had to live within 2 kilometres of the power plant. (Any further and you simply wouldn't get the oomph necessary to light your premises.) The only solution to the resistance problem was to boost the voltage – essentially kicking the electricity down the wire with greater force – but Edison considered this dangerous, working on the premise that if his customers came into direct contact with too high a voltage it might kill them, and that was bad for business.

You might have heard it said that 'it's not the volts, it's the amps that kill you,' so why was Edison so vexed about voltage? True, it is the current of electricity passing through your body that, if high enough, does the damage, but first it has to get into you – and luckily your skin resists electrical current. Increased voltages push through that resistance, kicking down the door to allow the current in (and if that current is north of 100 milliamps you're in trouble). Current is the assassin in your house, but voltage opens the window and lets them in.

The solution was the transformer – a device that could step voltages up or down (like the 'power packs' that sit on the wire between the plug socket and your computer, reducing the mains voltage to something that won't blow your laptop's mind). With their arrival, high voltages could be used to push electricity long distances through transmission cables suspended safely above ground but customers could receive a less dangerous 'stepped down' voltage via a local transformer.

Edison was still opposed, however. A high voltage was dangerous wherever it was he argued, as poor John Feeks had found out. He also had a less altruistic reason for opposing this new technology. The transformers of the day required alternating current (AC) to work* and Edison held many direct current (DC) patents that he wanted to profit from. He began a

---

* Transformers make use of the phenomenon that electrical current passing through a wire creates a magnetic field around that wire, and vice versa. You can use those magnetic fields as an intermediary to 'pass' electrical current from one wire to another. A coil of wire is wrapped around a bar of iron, thus generating a magnetic field within it. A second coil, wrapped around another part of the bar, has a current induced in it by that same magnetic field. Depending on how much wire is wound in each coil, you can increase or decrease a voltage. Why did early transformers require alternating current to work? Alternating currents, as the name suggests, flow first one way then the other in quick succession, creating constant changes of magnetic flux within the iron bar, which allowed the transformers to work continuously, rather than creating a brief 'one-off' transformation.

fierce and mean-spirited campaign against all things AC which became known as the 'War of Currents'. His tactics included having a number of animals (including an elephant called Topsy) publicly electrocuted with AC; lobbying government for a limit on power line voltages, and, even though he opposed capital punishment, helping to co-fund the first electric chair, on the understanding that it would be AC that finished off the condemned. Despite his best efforts, however, Edison lost the War of Currents, at the hands of two of his former employees. One was Nikola Tesla, a Serbian immigrant whose name is still synonymous with electrical innovation. The second was a man who became part of the inspiration for *Citizen Kane*.

On arriving in the States in 1881 Samuel Insull initially worked as Edison's personal assistant but soon became his second-in-command. However, it was his move to take charge of Chicago Edison in 1892 that led to him becoming known as 'the Henry Ford of the modern electricity industry'.

Like most cities at the time, Chicago had numerous local power plants serving the immediate vicinity with DC. Insull set about purchasing his incumbent rivals, turning his acquisitions into local substations filled with a new and powerful breed of AC transformer designed by Tesla. Able to withstand the high voltages necessary for citywide power transmission, Tesla's transformers allowed Insull to install huge generators at a few central locations and supply the whole of Chicago at reduced cost. By 1907, Insull and Tesla's AC network dominated the city. The now renamed Commonwealth Edison became known 'as one of the most progressive, and lowest cost, utilities in the world' and electricity companies across the country began to copy its model, with the result that 'electrical output from utility companies exploded from 5.9 million kilowatt-hours (kWh) in 1907 to 75.4 million kWh in 1927'.

Not everyone was happy. Insull's approach created, in effect, citywide monopolies and this worried people. On one hand, it was clear his entrepreneurial efforts (and those who followed in his footsteps) had reduced costs and brought electricity to more people. (In the first twenty years of the twentieth century, US electricity prices declined 55%.) On the other, policymakers worried that these market-dominating utilities might follow the example of the railroad industry – which, after a similar period of consolidation, saw the remaining firms go on to exploit their positions as sole suppliers, hiking rates while standards slipped and levels of service declined.

In an attempt not to repeat history, US policymakers adopted one of two approaches depending on the city or state. The first was municipal ownership. Cities might build their own energy generation facilities (or more usually bring existing private generators into public ownership). Without shareholders to keep happy, the theory was that citizens could continue to enjoy low rates without the risk of becoming easy meat for profiteering – a system of energy generation that, in theory at least, was by and for the people. The problem with this approach was the ever-present worry that public officials would at best not possess the right skills to oversee a complex energy system and, at worst, be corrupt or swayed by political motives that could lead to all the same abuses of power and poor performance that a corporate monopoly might be accused of.

The other approach was state regulation of electricity companies, where the government recognised utilities as 'natural monopolies' (services that, in order to operate at the lowest cost to the customer, are best provided by a single large organisation) but required them by law to adhere to certain conditions. (Today those conditions include providing a service to anyone who requests it – and can pay for it – usually at the regulator's approved prices, meeting safety and reliability standards and, to varying degrees, taking environmental and public health impacts into consideration.) Insull himself was an advocate of regulation, believing it legitimised monopolies and would increase public trust in them. As electricity generation spread around the globe, variations of these two broad approaches were combined at the same time as national energy grids sought to unify standards and ensure that demand anywhere in a country could be met even if local generation couldn't deliver (a requirement that had come into sharp focus for many nations trying to power their efforts during World War II).

But it turned out that neither approach was satisfactory. Too much power in too few hands, whether public or private, led to all the abuses and inefficiencies supporters of both sides feared – a problem that's still with us today. Whether it was the growth, subsequent sharp practice and collapse of US utility 'holding companies' in the 1920s and 1930s (by 1932 only eight holding companies controlled nearly three-quarters of the investor-owned utility business) or the failure of the UK's Central Electricity Generating Board to react well to the 1978 oil crisis, energy generation has been dogged by the 'natural monopoly' problem. Corruption, inefficiency, collective (if unintended) incompetence or profiteering are never far from the headlines. Insull himself died in disgrace with debts of $14 million after his own holding company folded during the Great Depression, taking the life savings of 600,000 citizens with it.

Later drives towards deregulation and privatisation in many nations were made with the hope that market forces would solve these problems – driving down prices and, through more competition, improving service levels, but true competition has barely emerged. The 'economies of scale' argument still seems to dominate, with a small selection of big players effectively creating a market-dominating cabal (now almost entirely owned by private investors).

But it doesn't have to be this way. Güssing's story points to a future where the natural monopoly isn't 'natural' any more, where the efficiencies, service levels and prices once considered only attainable by the big players can be achieved in local communities. And it's not only towns like Güssing that are beginning to understand this. The big energy companies are too and they don't like the implications. Some of them are even calling it 'Edison's revenge'.

Güssing is hard to get to. Its new-found prosperity has yet to translate into a train station or tourist accommodation, and there's precious little municipal transport, so I'm staying in neighbouring Jennersdorf, which has the benefit of a train stop and the charms of the modest Hotel Raffel. Jennersdorf is also home to Dr Joachim Tajmel, a local conservationist who doubles as an English-speaking guide to Güssing's many power plants. He arrives at the Raffel the next morning with his dog Kex. 'My assistant,' he jokes.

Normally the drive from Jennersdorf to Güssing takes about half an hour, but Dr Tajmel makes a slight detour so we pass through nearby Szentgotthárd. He's making a point. Szentgotthárd is in Hungary and when he was a boy this route would have been impossible. When the Iron Curtain descended, villages and towns on the western side of the border suddenly found themselves cut off from trading partners on the other side, decimating much of the local economy, one of the more recent chapters in the province's turbulent history.* Reaching the centre of Güssing, which is undeniably pretty (Joachim points out a particularly quaint monastery), our conversation turns to the local economy and how it's improved in recent years.

'People are staying now,' says Joachim. 'Today there are jobs here. Before, there wasn't much employment, a little agriculture, some forestry. Now it's industry that provides the work.'

'Who are the big employers today?' I ask.

'Good timing,' says Joachim. 'This', he indicates a building on our left, 'is a very important noodle factory. Yes, we produce a lot of noodles in Güssing!' I'm surprised but it turns out he means spaghetti and, whilst the Wolf Nudelin factory is notable for being 'the second biggest noodle factory in Austria', I'm more interested to learn it runs its operations almost entirely on power generated by its own sustainable 'biogas' plant, netting it the 2013 Austrian Energy Globe Award after impressing the judges with its 'CO_2-neutral noodle production'. Now Joachim points to a huge white building on the other side of the road, at least the length of a football pitch. 'This is Parador, a big flooring manufacturer. Yes, it's noodles and floors in Güssing!' Parador, it turns out, is one of two major flooring manufacturers who've been attracted to the town, the result of a proactive policy of corporate seduction by Peter Vadasz.

---

* Joachim tells me the story of his Hungarian-speaking ancestor – a local elementary teacher captured by the Russian army as they advanced on Austria-Hungary in 1916. For a year he languished in a Gulag in Siberia, but when Lenin deposed Nicholas II in the 1917 Russian Revolution, he was released, going on to run a restaurant in Vladivostok. 'He had a wife there,' Joachim tells me. 'Incredible!' But violence was never far away. The Russian Civil War, sparked by the Revolution, officially ended with the taking of Vladivostok, and during the fighting his wife was killed. Foreigners were banned from the city, as it was now the base for the Soviet Pacific Fleet, so 'he had to give up everything and came back on the last ship out. He returned here, but by then Burgenland was speaking German. So when he left he was teaching in a Hungarian-speaking school and when he came back he was teaching in a German-speaking one!'

'You have to contact CEOs, talk to them, offer them something to come here,' he'd told me. 'Nobody comes here because we have an old castle or a mad mayor!'

His pitch? Güssing could provide a ready supply of a flooring company's raw material (all those trees), but in addition it would pay for their sawdust and remnants and generate cheap energy from them. Joachim points out a long pipe by the side of the road. 'That's the tube through which we blow sawdust for burning at an electricity plant. Forty thousand tonnes a day!'

A few minutes later we've arrived at our first stop, a power plant I've specifically requested to see. It caused something of a stir when it opened, and its subsequent history is a telling lesson in what happens when new technologies challenge vested interests. The story of the plant may well turn out to be a condensed and prescient trailer of what will happen to the world energy market in the years to come. That it's still in operation, I learn later, is no mean feat – and testament to the resolve of Peter Vadasz and his partner in crime, the man who, when I mention his name, Joachim refers to as 'The Master'.

The plant is the world's first (are you ready?) Fast Internally Circulating Fluidised Bed Thermal Gasification plant. (There's a reason engineers don't go into marketing.) It may not be catchy, but it could catch on. The brainchild of Hermann Hofbauer (the meeting of whom Peter Vadasz claims was another of those 'best things that ever happened to me'), it's a new take on an old idea – that of turning mass into gas.

'Gasification' has been around for centuries. In 1609 Flemish mystic and chemist Jan Baptist van Helmont pondered how he could burn 62 pounds of charcoal but be left with only one pound of ash. Believing that matter could not be destroyed (this nearly a century and a half before the principle of 'conservation of mass' was first formalised by Mikhail Lomonosov), he concluded that the other 61 pounds of stuff had turned into a 'wild spirit', becoming something he called 'gas sylvestre'. With this thought he became the first person to understand that air is not a single gas but a combination, thus becoming the 'father' of 'pneumatic chemistry' (the name given to the study of gases' physical properties up until the early nineteenth century). Subsequent study disproved many of van Helmont's more mystical notions, but he still retains the credit for introducing the word 'gas' to the world.

Leap forward a hundred and eighty years and English colliery owner and inventor John Barber takes out a patent for the first gas turbine, a device which captures the 'gas sylvestre' given off by burning coal, cools it, mixes it with air and then ignites it to drive a simple combustion engine. It wasn't particularly efficient, but it was the earliest forebear of the workhorses that sit at the centre of modern gas-fired power stations and (via Frank Whittle) those things hanging off the wings of larger aircraft.

I'm looking at another descendant, the initial funds for which were raised by Peter: Güssing's Biomassekraftwerk. ('Kraftwerk' is German for 'power station'. In my experience few fans of the Düsseldorf synth-pop pioneers seem to know this.) The plant is bijou compared to your average power station (its footprint is not much bigger than a couple of basketball courts), but in nearly every other respect it's everything you'd expect, in that it's ugly and noisy. Girders, Heath Robinson pipework and various chimneys rumble and clang merrily. Parked in the shadow of the plant are three large trucks laden with woodchips. The scent of sap hangs in the air, rich and sticky.

'Bark,' says Joachim, which isn't an instruction to Kex, who is happily running about the place, but an explanation. 'This is waste bark, stripped from the branches and trunks of wood going to manufacturing.' Bark is the 'Biomass' in *Biomassekraftwerk*. 'Biomass' can refer to anything that's growing: trees, grass, crops, etc. 'It used to rot on the forest floor. Now we turn it into electricity.'

There's a lot of interest in biomass at the moment because it's classed as a renewable energy source. Sure, when you burn it biomass releases carbon, but that carbon can be reclaimed by planting new biomass to replace that we've incinerated. However, burn it without a replanting strategy and it'll contribute to rising carbon emissions as surely as burning coal (although coal generally releases more environment-damaging CO_2 than the biomass we use for energy).

Professor Hofbauer's wood-fed plant is important because it competes favourably with turbines fed by more energy-dense fossil fuels. His 'Fast Internally Circulating Fluidised Bed' shares the heat from the gasification process (heat for vaporising those incoming woodchips) with the combustion chamber (heat to burn the wood gas in the turbine in order to generate electricity) – a two-for-the-price-of-one approach that boosts efficiency. Actually it's a three-for-the-price-of-one system because the heat generated by the plant when it's creating electricity doesn't radiate away into the air but is pumped into the town's heating grid, warming homes and businesses, making *Biomassekraftwerk* what energy people call a 'combined heat and power' (CHP) plant.

Making use of that heat, if you've somewhere to send it, is a good idea. According to the coal industry's own figures, coal-fed power plants that don't find a use for their waste heat enjoy a measly efficiency of 33%. Bear in mind this is *most* coal plants. (The USA's Environmental Protection Agency reminds us that this state of affairs 'has remained virtually unchanged for four decades', meaning 'two-thirds of the energy in the fuel is lost'.)

But if you can make use of it, efficiencies can leap past 80%. The plant I'm looking at has an efficiency rating of 81.3%. Given it's running on woodchips, rather than a much more energy-dense fossil fuel, that's impressive, opening up the possibility that biomass can now compete very well with coal and gas as an energy source. *But* it's entirely dependent on the local heat grid to be viable. That grid, I will learn, is the key to Güssing's ability to prevail against the status quo.

It's not all about woodchips; it's also about bargaining ones.

Joachim introduces me to Gerald Weber, a researcher who's going to talk me through another surprising output from the plant: diesel. Yes, that's diesel made from woodchips. In fact, there's a petrol pump by the side of a white building next to the plant.

It sounds fantastical, but making liquid fuels out of wood (or coal) is also an old technology. You might have heard gasoline or diesel referred to as 'hydrocarbons', a reference to their chemical composition – part hydrogen, part carbon – and, when wood or coal are vaporised, the resulting gas is rich in both. In 1923, German chemists Franz Fischer and Hans Tropsch (with the help of colleague Helmut Pichler) worked out a way to recombine those gasified elements into liquid fuels. Gerald brings me inside to talk through the process – which hasn't changed a great deal since its invention.

'First we remove any impurities from the gas made by the power plant next door. They're not a problem in the turbine which burns them away, but they'll damage the Fischer–Tropsch reactor,' he says pointing to a shiny silver cylinder about 1.5 metres tall and 15 centimetres across. 'We pass the gas over a cobalt catalyst, which gives a chemical reaction – helping the hydrogen and carbon atoms recombine. They join into long chains on top of the cobalt – like trees growing up out of the forest floor. By altering the temperature and pressure we can build up molecules of different lengths. Shorter chains gives you jet fuel or gasoline; longer chains, you get diesel.'

Because the gas coming in is so thoroughly scrubbed of impurities, a key advantage of Fischer–Tropsch fuels is that they burn much cleaner than their fossil fuel cousins. In fact, right now the plant is creating high-quality (and premium price) 'blending products' for mixing with existing fuels – improving their performance and reducing the engine damage that impurities in regular fuels cause over time.

The Fischer–Tropsch process was an important tool for the Nazi war effort. With precious few oil reserves but plenty of coal, companies like IG Farben industrialised synthetic fuel production to keep the economy and army on the move during World War II. Thanks to its deep entanglement with the Nazis, and in particular its willing use of slave labour, IG Farben became the subject of the Nuremberg Trials, with senior figures indicted for war crimes. The firm was split up, one of its descendants being the chemical and pharmaceuticals giant Bayer. Following the war, South Africa (another coal-rich, but oil-poor nation with a repressive regime) invested heavily in the technology, building a series of huge plants that became a key tool in the apartheid government's ability to weather international sanctions. Elsewhere, however, the technology fell out of favour, largely due to the arrival of cheap Middle East oil in the 1950s and 1960s. It has enjoyed revivals during oil price shocks, and currently, as the world looks to wean itself off fossil fuels and oil prices rise again, there's renewed interest, which is keeping Gerald busy.

'It's a proven system,' says Gerald, 'and using biomass makes it more sustainable. You can't power a passenger plane on solar or using a battery. We still need liquid fuels. Here we can make them from the forest.' He smiles. 'It's pretty cool, no?'

As indeed is Joachim's next stop – which, I kid you not, is a huge fart generator. We all pass wind, and when we do, it's because microbes in our large intestine have been fermenting the food we have eaten, a by-product of which is gas – which has to go somewhere. Without these microbes, we would be unable to digest anything, and so we should accept the (on average) fifteen farts we do each day as *a good thing*. This same 'technology' can be employed on a much grander scale, and I'm looking at one such implementation: two circular green buildings with domed roofs. One is the 'fermenter' – a building of 1,500 smelly cubic metres, into which bacteria and biomass (grass, clover, corn, sunflowers, some unlucky rats) are poured, the subsequent reactions creating massive farts of biogas. The second building stores the results. This is no place for a crafty cigarette.

Joachim takes me into a nearby (incredibly hot and noisy) hut, where turbines turn the gas into electricity and heat for the 1,000 residents of nearby Strem. Next to the hut is a field of solar panels. 'Strem is investing in solar because it's a good way to generate cheap electricity, of course,' says Joachim as Kex sniffs what looks like some particularly pungent biomass. 'But everyone who invests also gets 4% interest on their money a year, which is a lot more than you'd get in the bank! Nobody is doing this for the environment; they're doing it for their wallets.'

Our next stop is Urbersdorf, a village of sixty or so households, five kilometres east of the main town. Joachim wants to show me one of the very first energy innovations in the area. 'It's quite old now,' he tells me.

We pull up outside a large barn, its roof covered with solar panels – the kind that heat hot water, as opposed to the silicon 'photovoltaics' I've recently seen in Strem.* Inside the barn Kex and I have fun climbing up and down the back-up system – a huge pile of woodchips that can be fed into a burner to heat the water during colder months. We move to the control room (passing enormous, gleaming hot-water tanks towering above us), where Joachim pulls up a real-time schematic of the system on a computer screen. Even though it's only April, the water, warmed by the spring sun, is a toasty 85°C, ready to be drawn on by the local residents connected to the barn via a super-local heating grid. He flips to another application, showing the consumption of each household in the system and their current bills.

'They pay about half the price they would with a utility,' says Joachim. Those cheap bills, I find out, include a levy for maintenance costs, explaining why the facility is still in fine working order after nearly two decades of operation.

'Well, everyone must be very happy with *that*,' I say.

Joachim shrugs. 'Not really.'

I'm surprised. What's not to like? But I've misunderstood him. It's not that they're unhappy, it's just that this micro hot water plant has been running for twenty years. Here, cheap, reliable, community-owned, renewable energy is nothing remarkable any more.

---

* The panels are filled with tiny tubes that carry a circulating 'thermal fluid', a substance which can draw a lot more heat out of the sun's rays than, say, a slice of bread. In fact, on really hot days the thermal fluid can reach temperatures well over the boiling point of water. That fluid is passed via a series of pipes through a water tank, delivering the heat it's collected into the surrounding H_2O before being circulated back up to the roof to repeat the process.

It's an attitude I come into contact with time and time again in the few days I spend in Güssing. From café owners to public officials, from taxi drivers to shop owners, there is a belief that there is absolutely nothing strange or unique about the way the town generates and distributes its own energy. 'Why would you do it any other way?' is a common refrain. That they find it unremarkable is, to me, completely the opposite.

'For us, it's normal. We have no other systems,' says Joachim. 'We use this. That's it.'

8 ENERGY TRILEMMA

'Status quo, you know, is Latin for "the mess we're in".' – RONALD REAGAN, ACTOR AND US PRESIDENT

'The Energy Trilemma' is a term, perhaps even a mantra, coined by the World Energy Council. The trilemma says it is almost impossible to deliver security of supply (the lights stay on), equitable access (everyone has the energy they need at a price they can afford) and environmental sustainability (we don't destroy the environment seeking to achieve the first two). It's the energy industry's job to do the best it can buffeted on the horns of this triple-headed beast. The trilemma is essentially an admission that the system we have cannot deliver the energy infrastructure we need for a prosperous, sustainable world.

Fossil fuel lobbyists are quick to point out, for instance, that you can't have 'security of supply' with renewable energy sources because of the 'intermittency problem' I'd investigated in Birmingham with Yulong Ding. A society relying on renewable energy would regularly face blackouts, they argue – and until we have mass adoption of utility grade batteries they've got a point. Oil, coal and gas, by contrast, are their own battery. To release their energy we need only set fire to them.

Environmentalists counter that continuing to burn fossil fuels will lead to catastrophic climate change, a disaster for the planet and by extension our future prosperity – no 'environmental sustainability'. As Gaylord Nelson (former US Senator and founder of Earth Day) said, 'The economy is a wholly owned subsidiary of the natural environment, not the other way around.' Both groups therefore bemoan the subsidies given to the other, as governments try to satisfy the third need expressed in the trilemma: affordable and thus equitable access to energy for their citizens.

So how do we solve the trilemma? Green energy activists can't deny they have benefited from the energy dividend of fossil fuels. Every defender of fossil fuels knows that Gaylord Nelson was right. For those of us not in the energy industry it's almost impossible to work out the true picture. But, in examining subsidies, the key tool used to try and satisfy the demand of 'equitable access', some light begins to shine in the darkness. So, ahead of my meeting with 'The Master' tomorrow, I'm in my hotel room boning up on the labyrinthine world of energy subsidies.

You've probably witnessed renewables subsidies being debated in the media, the most popular battleground being 'feed-in tariffs' (or 'green taxes', depending on your viewpoint) – guaranteed, and relatively generous, payments to renewable generators for their energy. Indeed, Güssing's *Biomassekraftwerk* qualifies for a generous feed-in tariff for the electricity it generates. Critics argue that this encourages the use of inefficient, intermittent and expensive forms of generation. Environmentalists and the renewables industry counter by pointing out that fossil fuels *also* benefit from subsidies, which are justified using the same argument – that people need access to energy and it's too expensive without some form of support. The fossil fuel lobby respond with the argument that, while that's true, if you want to help the world's poor it makes more sense to subsidise fuel sources that are generally more available, reliable 24/7 and cheaper (i.e. fossil fuels). Your middle-class and pricey green agenda isn't much use to those who desperately need access to energy to lift themselves out of poverty.

Both sides quote figures on the outrageous amounts the other is getting but, of course, those figures vary widely. The International Energy Agency (IEA), for instance, estimates that fossil fuel subsidies are about $500 billion worldwide while researchers at the International Monetary Fund think the figure is nearly $5 *trillion* – ten times bigger. What gives?

Actually, both are right, depending on what you consider constitutes a subsidy – and, by unpicking *that*, the world energy market, the trilemma which defines it and the arguments for different energy systems become easier to understand. Güssing's transformation also becomes easier to assess. I need to be sure: has the town broken the trilemma, or is it a heavily subsidised curio?

The simplest (and, some argue, most honest) way to measure subsidies is to look at direct monetary support from government that benefits either consumers or producers of energy. The most extreme example of a 'consumer' cash subsidy comes from Venezuela, whose government has a history of subsidising gasoline prices so vigorously that a gallon of petrol can cost as little as two cents. An example of a 'producer' cash subsidy might be one of those 'feed-in tariffs' which guarantee that producers of certain types of energy (often solar) will be paid over the market rate for their energy.* Some people

---

* Some countries, including the UK, get past the problem of having to pay this money to producers by asking utility companies to pay the tariff for them. Those companies pass the cost on to customers through slightly increased electricity bills. Depending on who you talk to, this is either a sensible way of using an existing energy marketplace to bypass the expense of collecting and redistributing tax, or an unfair and corrupt system which arbitrarily penalises utility companies and/or their customers. It also muddies the waters as to whether a feed-in tariff counts as a 'direct' subsidy or not, because in this model the government isn't paying anyone directly. For my UK analysis I'm putting feed-in tariffs, even if they're administered via the existing market, in the 'direct' category.

also include government support for energy research and development as 'direct' subsidies too. If we take this definition of subsidy, how do the figures stack up for fossil fuels versus renewables?

Let's take the United States. In 2013, government subsidies for coal, natural gas and oil totalled $3.4 billion, whereas renewables enjoyed support of over four times that at just over $15 billion. If you then take into account that green energy produced a little over 10% of the energy used in the USA that year, then, per unit of energy, renewables enjoyed *forty* times the subsidy of fossil fuels. The UK has been even more generous. A 2011 report on subsidies commissioned by Parliament's Environmental Audit Committee estimated that, while 'direct' subsidies for renewables totalled about £2.4 billion that year, incentives for producing oil, gas and coal were much smaller, at £284 million. Per unit of energy generated, renewables received over 250 times the direct subsidies fossil fuels did.[*]

Worldwide renewables tend to get favourable subsidies per-unit-of-energy produced, but, almost inevitably, the absolute figures for fossil fuel subsidies are much higher – 'around six times the level of support to renewable energy', says the IEA. This, say fossil fuel advocates, is good for the poor – giving cheaper energy access to those that most need it. That argument, however, doesn't stand up to much scrutiny. The richer you get, the more energy you use, so fossil fuel subsidies disproportionately benefit the wealthier members of a society.

---

[*] It's also worth bearing in mind that in the UK fossil fuel subsidies were largely funded from taxes imposed on the fossil fuel companies in the first place, notably in the form of the Petroleum Revenue Tax (PRT), which levies extra taxes over and above Corporation Tax on the profits of certain historical oil fields. Is a subsidy really a subsidy when it's simply giving you back tax you've been specially singled out for? This may be one reason the Chancellor effectively axed PRT in his 2016 spring budget.

As Maria van der Hoeven, the IEA's Executive Director, points out:

> 'Let us be clear: fossil fuel subsidies are an extremely inefficient means of achieving their stated objective, which is typically to help the poor. IEA analysis indicates that only 8 of the money spent reaches the poorest 20 of the population. Other direct forms of welfare support would cost much less.'

But of course the story doesn't end there. Next we need to consider 'indirect' subsidies, such as tax relief – a well-worn tool for encouraging us to buy things our governments think are important or socially useful. For instance, the standard sales tax in the UK (VAT) is 20%, but many activities are exempt or enjoy a lower rate (from small business sales to incontinence products). When it comes to energy, VAT is levied at a quarter of the standard rate, at 5%. If we include the 15% non-levied tax in the definition of 'a subsidy', the picture of who gets what changes considerably. Fossil fuels benefit the most because they're the dominant part of the energy mix. And again it's argued that this, rather than benefiting the poor, skews the savings to those who use the most energy (the rich). That said, renewables often still get a good deal. In the UK, for instance, even with the lower VAT rate on energy taken into account, they were still subsidised, per-unit-of-energy, thirteen times more than fossil fuels were in 2011.

Now we come to the most controversial component of subsidy, which if you include it gives you the $5 trillion figure the IMF quotes for fossil fuel support – i.e. the additional costs *society has to bear* as the result of using any particular fuel. If we include this component, the picture reverses entirely, with the subsidies (if you accept this wider definition) given to fossil fuels dwarfing those enjoyed to renewables.

In the parlance of economists, these extra costs are called 'negative externalities' – stuff one person or organisation does which adversely affects another person or organisation without them asking for it. Airports, for instance, visit the nuisance of noise pollution on nearby residents, but airline ticket prices don't include a contribution to soundproof the neighbourhood. A resident wanting to deal with the unwanted noise will have to find their own solution at their own cost – a 'negative externality'. Fossil fuels, like many other products, come with negative externalities – and they're humdingers.[*]

First up is air pollution. My time investigating Peter Dearman's liquid-air engine has already brought the pollution problem to my attention, but digging a little deeper I come to realise what a massive challenge it is. The figures are astounding. The World Health Organisation estimates air pollution, largely a result of fossil fuels being burned, is responsible for a shocking *one in eight* deaths worldwide each year, the vast majority in the rapidly urbanising economies of Asia and Africa. That makes it arguably *the biggest killer in the world* (in an unsavoury battle for the accolade with heart disease and stroke).

A 2013 study by the US Environmental Protection Agency concluded that the effects of air pollution could be stripping up to 6% (nearly $900 billion) off the national GDP. This came on the back of a report by the Health and Environment Alliance

---

[*] Externalities can be positive as well as negative. A good example is vaccination. The person receiving the vaccination gets a clear benefit, i.e. protecting themselves from disease. But the great thing about vaccination is that it can be good for an entire population, even those who aren't immunised. If enough people get vaccinated, infections find it harder to spread. This is called 'herd immunity'. So even if you're not vaccinated you've been offered some protection by those who have – a 'positive externality'. There are economic benefits too, in that a largely vaccinated population keeps people, even the unvaccinated, alive to contribute to society and the economy. So, vaccination's positive externalities are as much cold and economic as they are compassionate and medical.

(an EU-funded NGO independent of any political party or commercial interests) that concluded European coal plants came with a hidden €42.8 billion health bill attached each year. In the UK, the government estimates that air pollution robs the population of 340,000 years of life annually (the equivalent of 29,000 deaths). In China, air pollution kills more people than smoking.

The next externality to consider is climate change. Investors and business are beginning to wake up to the reality that climate change costs money. An attempt to quantify the economic risks to the United States, backed and supported by heavyweight capitalists like Michael Bloomberg, Henry Paulson and Tom Steyer,* concluded, amongst other things, that repairing the damage wrought by climate change-induced extreme weather will soon cost $35 billion a year, that sea level rise could sink $106 billion of coastal property by 2050, and that 'some states in the Southeast, lower Great Plains, and MidWest risk up to a 50% to 70% loss in average annual crop yield'. As Johan Rockström, Executive Director of the Stockholm Resilience Centre, says, 'You may have noticed the environment is starting to send back invoices.'

It's by factoring in these externalities that the IMF arrived at its figure of $5 trillion subsidy figure for fossil fuels. This, like everything in energy, is controversial. Is an unpaid-for externality really a subsidy? Some people (usually those working for large fossil fuel companies) argue it's stretching the definition too far. This wrangling over definitions doesn't, however, change the fact

---

* Bloomberg: business magnate, ex-mayor of New York, and tenth richest man in America. Paulson: ex-CEO of Goldman Sachs and former US Secretary of the Treasury. Steyer: one of the world's most successful hedge fund managers who famously turned the $8 million he started with in 1986 into $30 billion by the time he left the fund in 2012.

that our dominant source of energy is killing millions of us via pollution already and, if carbon emissions aren't brought down dramatically, will visit on us an environmental, humanitarian and economic crisis of unprecedented proportions. In this context fossil fuels don't seem like great value for money, not now there are genuine alternatives, as my visit to Güssing is demonstrating.

I imagine a future where a grandchild might come to me and say, 'So let me get this right, Grandad … when you were younger people used to dig up coal and oil from millions of years down, and that was dangerous and expensive, and destroyed the places (and upset the people) where it happened? Then you'd burn it in big buildings to create electricity, or pour it into engines, but those buildings and engines would lose most of the energy as waste heat? And people used to go to war for the stuff in the ground because everyone wanted it so much, but when you burnt it, it killed more people than all those wars put together because of pollution? Oh, and burning it also created the biggest threat to mankind in history, because it was heating up the world too much? Did I get that right?'

And I'd have to say, 'Yes, that's about it.'

'And the way you made food …'

'Why don't you go chat to your grandmother?'

Tomorrow I'll meet the man who stood overlooking the battlements of Burg Güssing with Peter and whose love of his hometown's fresh air and surrounding countryside convinced him it was time for a revolution.

I'm in a nondescript meeting room in the European Centre for Renewable Energy, the HQ from which all of Güssing's energy projects are coordinated, of which I've seen only a small selection so far. I could be anywhere, except for the fact that along one wall I count over twenty energy and sustainability awards from all over the world. One, awarded in Japan, relates to the biomass plant I saw yesterday: 'In recognition of your outstanding contribution to the resolution of global environmental problems and to the creation of a sustainable future'. Japan has been interested in Güssing for a while, exploring options to reduce the nation's dependence on energy imports and replace domestic nuclear generation that's become increasingly unpopular after the meltdown of the Fukushima power plant triggered by 2011's catastrophic tsunami. The award bears the name 'Reinhard Koch': the man Joachim called 'The Master'.

The first thing you notice about Reinhard ('Reinnie' to his friends) is his height. At six-foot-nine, he is, by most standards, *freakishly* tall. When he walks into the room I immediately wonder why he didn't become a basketball player instead of an engineer. He's instantly likeable, confident without being arrogant, questioning but not aggressive, exact in expression without pedantry. A smart kid from a local middle-class family, he studied electrical engineering but, like almost everyone else it seems, had to move to find work.

'Did you know the capital of Burgenland is Chicago?' he asks. My look of confusion prompts an explanation.

'Between World War I and World War II, 90% of the population left the province, lots to America. Today there are more than half a million Burgenlanders in Chicago!' That's considerably more than the province's current population of less than 300,000 and has created a few problems for the collection of local biomass. 'A lot of the owners of forests and agriculture fields are in the States. Many didn't even know they owned it. It was left to them, but they've never been here.'

Born in the 1960s, Reinhard didn't go as far as Chicago, but like many of his generation did leave town – 'For me and my friends, we knew: you finish school, then we have to go to Vienna. Boom!' Most weekends, however, he would return to his hometown of Strem, not only because he loved the countryside and how it changed with the seasons, but also to indulge his passion for – aha! – basketball, being for many years a star player on the Güssing Knights (a club he now owns). A man of clear ambition, he soon progressed, both in his day job and the sport, co-founding an engineering practice in the capital that won major contracts while simultaneously playing for his new club, the nearby Klosterneuburg Dukes, one of Austria's most successful teams. When he wasn't scoring points for the Dukes, he was part of the Austrian national squad.

'That was very important for my thinking, especially the trips to Asia,' he tells me. 'I thought I had to take something of Asian philosophies and bring them back to Austria.'

The sport also introduced Reinhard to the love of his life in the form of a Viennese-born basketball player whom he married. By 28, he was a successful engineer, an internationally recognised sportsman and a father. This last development changed everything. 'I said to my wife, "I want to go back home."'

Despite travelling the world, he remained convinced Güssing district was special. 'I know the whole world, but this region …' he points out the window. 'It's a really lovely region. Summer, winter, spring and autumn are all beautiful.' He felt strongly his children should grow up surrounded by the beauty and open spaces of his youth, rather than the urban sprawl of the capital. But he needed a job and a lack of those in Güssing was the reason he'd left in the first place.

'I asked the former mayor, the mayor before Peter, if he had anything,' recalls Reinhard.

He was lucky, securing a position in one of the few ways it was still possible in Güssing; replacing someone who'd recently died – in this case, the manager of the municipal water plant. With typical ambition he told the mayor that was fine to begin with but he'd need 'a bigger job than that'.

'We decided I would be responsible for the wastewater system of the whole region.'

Not long after starting, Reinhard approached his boss with a further request. 'Mr Mayor, I want *also* to be responsible for the energy system.'

Reinhard's ambition was born of his belief that a town should provide *all* the basic utilities for its citizens. In Güssing the local administration already oversaw drinking water and waste water, but not energy – this being in the hands of private energy providers. But, with new technologies and a different financial model, Reinhard was convinced a new approach could outperform the incumbent centralised, privately owned system, while keeping money in the local economy.

Of course, most people thought he was crazy. But Reinhard had two things in his favour. The first was the election of a new mayor, Peter Vadasz, who shared Reinhard's philosophy. The second was that Reinhard had become, he freely admits, 'a local hero'. People trusted him. In a town as down on its luck as Güssing was back then, a national sporting star was one of the few lightning rods for aspiration. At a time when nearly anyone who could get out of town did so, Reinhard – arguably its most talented son – was coming back. When it was time to convince people to try something different, his pedigree as a sportsman was crucial. People knew that here was a man unafraid to work hard and reach for the top – but, just as importantly, he usually delivered a good result for the team he was on. Reinhard was seen, both literally and figuratively, as a safe pair of hands.

Early projects included the small solar plant in Urbersdorf I saw yesterday, followed three years later by a full-blown district heating network for Güssing itself. Peter Vadasz convinced the town council that all public buildings should become customers of the new system, giving the project a sustainable economic underpinning. This, it turns out, was a masterstroke, allowing Güssing to create a town-wide network that would act as a shield against a series of attacks that were soon to follow.

Has Güssing found a way past the energy trilemma, I wonder? I've been impressed by what I've seen so far, but there's no getting away from the fact that the town has enjoyed a generous influx of EU grants. Estimates vary, but my best guess is between €25 and €30 million, a figure Reinhard seems to think is in the right ballpark. Does Reinhard's energy philosophy make economic sense? Or is it propped up by subsidy? I can't leave without finding out, but, however I couch the question, I'm about to ask the man in front of me if his whole philosophy amounts to an expensive hobby funded by the EU. When I do he looks almost pained. You can tell he's been through this argument more times than he cares to count.

'Getting money from the EU is great, yes … but this idea, this philosophy, *cannot* work unless you can prove the ownership and the economics work locally. So, in Urbersdorf, a lot of the capital was actually a bank loan, which is now paid off. When it comes to the district heating grid, people were already buying oil for their boilers so they could be asked to pay to connect to the district heating. You need to get as much money from inside the community as possible. But now you keep that money *in* the local economy. You have to prove that these systems can fund themselves – otherwise, it's pointless, it's not sustainable.'

'So, how does it work out in the long run?' I ask. 'What's the payback?'

'In total the investment in energy projects for the region has been about €80 million. About thirty per cent of that was grants, yes. But the rest was investment by the people. And the economic dividend?' He smiles. 'Our best guess is ten times the cost. About €800 million.' That's largely down to a virtuous circle of money staying in the local economy, stimulating economic growth that in turn brings more businesses to the town, who employ local people, who have more money to spend ... and so on. 'One of my sons now plays for The Knights and he has a teammate who wants to be a lawyer. Today he can do that here in Güssing.'

But as Güssing's fortunes began to improve, some people were less than happy. In the early days of *Biomassekraftwerk*, Reinhard did a deal with the province's energy utility, who were interested in Hermann Hofbauer's approach to gasification. They saw the plant as a way to learn about a potentially profitable new technology for their own business and Reinhard was happy to share operational and research costs. But, as the project progressed, Reinhard tells me, the utility began to see that coupled with the heating grid, the plant was part of an energy generation system that didn't require them. Trying to head off the threat, he tells the story of how they offered to buy the district heating system from the town, a strategy to regain control of the local energy landscape. Except the town wasn't selling.

'So, they said, "OK then, we will kill your gasification plant",' recalls Reinhard. 'They pulled out all their money. Then they went to our bank and told them to cut all our lines of credit, overnight. Now, understand this. The bank manager is a personal friend of Peter, the mayor, but still he does it!' Reinhard's eyes narrow a little, remembering the betrayal. 'We asked him, "Are you *crazy*?" and he said, "I'm sorry, but the energy company is my biggest customer, I have to do it." Then he told us we had to repay all our loans by the next day.'

'What did you do?' Reinhard gives me a look that twinkles with mischief but hints at a deep resolve.

'I found another investor.'

'How?'

He laughs. 'Quickly!'

Today the plant is owned entirely by the investor Reinhard called on (a personal contact), who also went on to suffer from a touch of megalomania, trying a few years later to buy the district heating system too. He met with the same response as the utility had. The grid was not for sale. The plant, Reinhard had made sure, was dependent on the town, not the other way round. Without a market for its waste heat, it wouldn't be viable.

'He makes a profit but he can only do that because we buy his heat. It's also the case he has to buy local biomass – if he ships it in from too far away, it's too expensive and he makes a loss. So, again, we keep money circulating locally, rather than flowing out of the town.' It's what a basketball coach might call a 'zone defence' – the heating grid is inaccessible to the enemy. Reinhard has controlled the court. 'The system is designed to support the townspeople. They are in charge. He's the owner of the plant, yes, but he's *not the owner of the system*.'

I take the opportunity to ask Reinhard what advice he'd give himself if he could go back twenty years.

'I'd have accepted fewer grants. If you want to build an energy system for the people, try as much as possible to build it with their money, not someone else's. The bank's money, the government's money – it always comes with strings. Today, we are doing everything without grants. On Saturday, we will open the new photovoltaic plant in my hometown of Strem that you saw yesterday. When the mayor announced it and said it would cost €100,000, the people put up that money in three hours. That is our way.'

'They trust the system now?' I ask.

Reinhard's answer is a slam dunk.

'They trust the system,' he says, 'because they own the system.'

9 MAKE WAY FOR THE ENERNET

'To change something, build a new model that makes the existing model obsolete.'

– R. BUCKMINSTER FULLER, ARCHITECT AND SYSTEMS THEORIST

When Samuel Insull kick-started a US energy revolution at the dawn of the twentieth century, his motivations were to take more energy to more people more cheaply, and in doing so he helped put in place a key foundation of a world-dominating USA economy. Insull did this by concentrating power, both physically and economically. Reinhard has proved, at least in Güssing, that he can achieve the same results using a distributed approach where the people own most of the assets. Two models separated by a century with the same ambitions but very different power structures.

There's a battle coming. In fact, it's already started.

David Crane, one of the electricity industry's most outspoken CEOs (he runs NRG Energy, a huge wholesale energy generator with 50,000 megawatts of power generating capacity 'capable of supporting nearly a third of the US population'), told the *Atlantic* that a future where communities were much more involved in their own energy generation would be 'utterly destructive of the utility model that we now have'. When he thinks about competitors, he admits he gives very little thought to the traditional players. 'Utility executives are usually the antithesis of visionaries,' he says, which I suspect means he's not invited to a lot of industry parties. Instead, he believes the energy market will become a 'free-for-all'. He isn't alone in his analysis.

A 2014 report by UBS advised the bank's investment clients 'solar systems and batteries will be disruptive technologies for the electricity system' and 'generation-heavy utilities will be relative losers, as large-scale power stations will hardly fit into the new, decentralised electricity world'. In their hearts the utilities know it too. The Edison Electric Institute, the association representing all the privately owned electric utilities in the USA, has concluded that an emergence of distributed energy technologies (in particular solar) will be largely responsible for 'declining utility revenues, increasing costs, and lower profitability potential', drawing comparisons with how the emergence of mobile phones spelled a decline in revenues for fixed line telephony. Perhaps this is why the Institute is lobbying for home solar owners to pay extra levies for their grid connection, now that they are energy suppliers – and investing millions each year in anti-solar PR.

Let's be clear here. The incumbents are looking at potential economic carnage, and they don't like it.

Understanding the fight to control Güssing's heating grid has helped me grasp a wider truth: that the key battleground in energy is not generating capacity (important though that is), but access to energy markets. While Güssing was able to win the battle when it came to heat distribution, it still has to play by the existing rules when it comes to electricity. The town remains dependent on the Austrian Power Grid – the privately owned (but state regulated) distribution system for the entire nation. Energy that isn't used immediately or stored must be sold to the local utility company. This of course provides income, but the town still has to deal with a middleman setting the prices.

Before setting out on my trip, I'd attended a community energy conference in Manchester called, appropriately, 'Powering up the North', as did Dr Jeff Hardy from the UK energy regulator Ofgem. Jeff is Ofgem's Head of Sustainable Energy Futures, looking at 'non-traditional business models' – one of the people, from the government side, who is trying to work out how the UK's energy system can transition from a system dominated by large, but largely disliked, natural monopolies to a new, more inclusive lower-carbon alternative. As the government's man in the room, I felt for Jeff who was, in no small measure, set upon by a crowd of vocal and passionate community energy leaders. Their frustrations were many, but foremost was an anger that they couldn't sell their energy directly to local residents at a price they chose, instead being forced to sell it to a utility at a mandated price, then to buy it back at another price set by a middleman.

Building their own local electricity grids is one option for these communities, but this brings with it all the problems the grid was originally designed to solve, most notably ensuring there's electricity even if your local provider goes offline or can't meet demand. An extended period of still weather for a local grid fed only by a nearby wind farm would see the lights go out once any local storage facilities had been exhausted. Similarly, local solar generation might suffice in the summer months but leave residents blacked out in the winter. A method of importing energy to areas that have a shortage of supply, or exporting it when they have a surplus, is *exactly* what the existing grid was designed to do. Heating grids *have* to be local because the mechanics don't work at a national scale, but creating separate local electricity grids, whilst intuitively attractive, comes with a whole bunch of unpalatable compromises.

It makes more sense, therefore, to give communities generating their own energy direct access to the existing grid – allowing them to buy and sell freely to their own residents and, when needed, other communities. But of course those who own the system (usually energy companies) are less than keen. Why would they open up their key mechanism of control to allow a flood of competitors less concerned about profit to enter the fray?

Luckily for citizens, whilst national grids are largely owned and operated by private companies, those companies are still regulated by governments, who are slowly coming around to the idea that the grid needs to open up to community renewables. That, however, doesn't stop their representatives, like poor Jeff Hardy, getting a kicking every time they go out in public. Of course this distributed energy future needs a solution to the storage problem, but with inventions like Yulong's air battery (and the general downward trend in battery prices) the solution to that particular gremlin is already on the horizon.

'Why can't we be more like Germany?' asked one attendee. They were referring to Germany's world-leading stance in giving its communities access to the energy marketplace. Earlier in the day, the conference had heard Caroline Julian, Head of Policy and Society at UK thinktank ResPublica, recount her tour of Germany's energy system, the written report of which I immediately downloaded. What struck me most from reading it was how, thanks to a tradition of strong municipal government (localism is a cornerstone of German politics), grid ownership in the country is distributed among many players and is therefore much more malleable. In fact, Germany has 888 local grid operators. This absence of market-dominating players is one of the reasons why, by the end of 2012, '190 German communities had been successful in bidding to run their local electricity distribution grid', writes Caroline. In addition, 'almost half of all electricity supply companies are owned by local government, communities and small businesses, with many *increasingly competing privately-owned utilities out of the market*' (my emphasis).

Another sentence caught my eye: 'Many of these emerging community-owned grid operators and suppliers are not only offering cheaper tariffs than their competitors, but are seeking and fuelling the prosperity of their locality' – just like Güssing. Those smaller, and in some cases cheaper, players are almost exclusively generating energy from renewables, which on sunny days have been known to provide over 70% of Germany's electricity. (On average renewables account for roughly 30% of the country's electricity each year.) By 2013, 47% of renewable electricity generation was already owned by citizens, with only 5% in the hands of Germany's big four energy providers and the rest split between industry, private investors and banks. If this trend continues, it brings with it the possibility of oversupply, which tends to have a downwards effect on prices. In fact, both Germany and the UK, on particularly sunny days, have seen wholesale electricity price go *negative*.* Cheap energy is not a problem if you're a community looking for a local economic boost, but it might be a worry if you're a big energy company who needs customers to satisfy its shareholders' demands for healthy profits.

It's early days, but already the savvier utilities are recognising that they'll be squeezed out of the market if they don't change course. In November 2014, seeing a real threat to its business, Germany's

---

* This is largely down to the inflexibility of the current system. When renewables start to generate large amounts of energy, you need to take other power plants offline, or increase demand. But shutting down coal and nuclear power plants (only to power them up a few hours later) is neither practical nor cheap. Bizarrely it makes economic sense to increase demand (by paying people to use energy) instead.

biggest energy supplier E.ON announced its intention to split itself in two, creating a fossil fuel company and a separate renewables business. Interested by E.ON's proposed split, I interviewed, off the record, someone who'd helped author an internal analysis of the company's financial future. They told me, 'we'd just refitted our German coal plants to run for another twenty years and our post-investment reviews showed we were set to make a loss over their lifetime. I remember being in meetings with the people running the power plants who were getting a complete grilling over their numbers, as our asset managers realised that, even though the plants were performing as well as they could, renewables were sneaking up on them.' E.ON hopes to firmly rebrand itself as a renewable energy company that, after lagging behind community investment in wind and solar, is keen to get seriously into the game. 'The new company has been given an *enormous* pot of money,' my contact told me. 'They want to build and buy market-dominating amounts of solar, wind, biomass plants, the lot.'

That E.ON is attempting to ring-fence the risks associated with fossil fuels while tooling itself up to compete in a market where the grid is, as David Crane of NRG puts it, 'a free-for-all', is illuminating. It also looks like a smart move, given that some months after E.ON's arrangement the German government, as part of its drive to decarbonise its economy, decided it was going to annex 2.7 gigawatts of coal-fired power plants from the grid. 'The affected power plants will not be allowed to sell electricity on the normal energy market,' said Energy Minister Sigmar Gabriel. It's a move only made possible because renewable sources from the community can take up the slack. Here comes the sun.

This journey to a distributed energy system has been called the 'Thousand Flowers pathway' and it presents enormous challenges to everyone involved, inviting a radical rethink that will come as an unwelcome shock to many of the existing players. Germany leads the way, thanks in part to a strong history of localism, but its municipal structures are not paralleled in many other nations.

How, then, can a country like the UK move from 'the big six' energy suppliers to a community of 60,000 local generators? How might the US energy system reshape itself to embrace energy localism? And how should we interpret the purchase of grid systems in Australia, Brazil, Portugal and the Philippines by the State Grid Corporation of China (the largest electricity company in the world), which understands well the power that comes when you control the means of distribution?

Is it possible to embrace renewable generation and meet worldwide demand, all while opening up energy markets for many more players? Like many questions these days, you can find the answer on the Internet.

'Yeah,' says James, with a chuckle. 'I suppose it is pretty hardcore.'

It's not a bad assessment. After all, James Johnston wants to totally reboot the world's energy distribution networks. Not yet 30, his eyes sparkle with an unrestrained youthful ambition, something he'll need to hang on to in the years to come. It's a few months before my trip to Austria and I'm in a noisy and achingly trendy coffee house in London's Shoreditch. James has cycled from his nearby office, the bijou headquarters of Open Utility to meet me.

'So basically, the question I'm asking, and Open Utility are trying to answer is, "Why can't you access energy on the grid the way you access information on the Internet?" When I access a webpage, I'm asking for information from a specific source and it comes to my browser. It could be any kind of information, it could be coming from anywhere in the world. Why can't I do the same with energy? Why can't I request power from a specific source over the electricity grid?'

As a PhD researcher studying local micro-grids, James found himself watching a lecture by Internet entrepreneur Bob Metcalfe at the Singularity University, the thinktank/start-up incubator/academic institution established by futurists Ray Kurzweil and Peter Diamandis to 'educate, inspire and empower leaders to apply exponential technologies to address humanity's grand challenges'. Bob's something of a legendary figure in technology and business circles. He co-invented Ethernet, the foundation technology for most computer networks today. (If you've ever seen a movie where someone is trying to shut down a network by yanking cables out of the back of computers, it's usually fistfuls of Ethernet cables they're pulling at). He commercialised his invention to great effect, creating 3Com (acquired by Hewlett-Packard for a cool $2.7 billion in 2009). He's also the originator of Metcalfe's Law, which famously states the usefulness of a network rises exponentially as more people or devices are attached to it – one of the foundational truisms of our connected age.*

As such he's rightly considered a key figure in the birth of the Internet age. He's opinionated, but never less than entertaining, even when he's spectacularly wrong. In 1995, in his then weekly column for *Infoworld* he incorrectly predicted that the Internet would 'catastrophically collapse' the following year, and if it didn't, he'd eat his words. As a result he was required to submit to a 'highly theatrical public penance: In front of an audience,

---

* The strict definition of the rule is that the value of a network grows by the *square* of the size of the network. So whilst an unattached computer has a value of one ($1^2=1$) a network of five computers has a value of twenty-five ($5^2=25$) whilst a network of thousand computers has a value of a million ($1,000^2 = 1,000,000$)… and so on.

he put that particular column into a blender, poured in some water, and proceeded to eat the resulting frappe with a spoon.'

After leaving 3Com, Bob was inspired by MIT president Susan Hockfield to think about energy, and began to wonder if the lessons he'd learnt helping to build the Internet could usefully be applied to creating a better energy system. After all, the Internet is really the most spectacular network ever imagined. It readily attaches technologies of all shapes and sizes, new and old – and allows communication between them. It can upgrade itself in a piecemeal fashion rather than requiring a complete rebuild, and it doesn't have to be shut down when maintenance or upgrades are required. It can distribute what it's asked to quickly, to wherever it's needed. It's fault-tolerant. It serves billions. And, while there are a few big players on the Internet, they don't own it. It is therefore, says Metcalfe, not a bad model for how we could run an energy system.

When challenged by energy professionals as to why he was getting in on their game, he recalls, 'I simply explained to the energy people that they'd had the problem for a hundred years and *hadn't* solved it, so they should step aside and let the Internet people [have a go].'

How did the Internet take over the world? The short answer is because the Internet isn't one network, but four. Early in its history some computer visionaries (most notably Vint Cerf and Bob Kahn) were trying to answer the question, 'How can we reliably get data that lives on a computer on *this* network to another computer on *that* network when both might be using different hardware and software?' – a problem then called the 'Inter-Net problem'. Their answer was a set of 'software protocols'

(think 'rules of engagement' for computers) with the decidedly unsnappy title of 'Transmission Control Protocol/Internet Protocol', or TCP/IP for short. (You may have come across this unfriendly couplet when you've been battling with your Internet provider or heard someone talking about 'IP addresses'.) TCP/IP splits computer networks into four 'layers' that, though interlinked, operate independently. At the bottom is the layer that defines how data is physically sent through the network, whether that's coaxial cable, optical fibre or copper wire. Next is the 'Internet layer' that splits data into discrete packets to be sent through those wires.* The next layer is called the transport layer, a 'handshaking protocol' which sets up agreements that allow two different computers to talk to each other. Finally, the application layer describes how the applications you use (e.g. your browser or email client) access the data arriving at your device and prepare their own data to be sent back across the network.

Metcalfe says the creation of TCP/IP (and its slightly more geeky cousin, the 'ISO seven-layer model', which is a thinner slicing of the same) 'was the most important decision in the creation of the Internet'. By splitting the 'Inter-Net problem' into a set of layered protocols with agreed ways of passing their results up or down to the next layer, different experts within each layer could apply themselves to solving problems without having to worry about what was happening in the other layers.

---

* On the Internet there is no fixed circuit between origin and destination along which data can pass. Instead, data is split up into numbered 'packets', which are sent into the wilds of the 'Net to find their own way to the destination. As each packet passes through any of the routing points on the network, it says, 'Hey, I'm trying to get to B, do you know where B is?' and one of three answers will come back: 'Yes! *I* am B,' 'Yes, B is over there' or 'No, but I'm sending you to another machine who might know where B is.' (When I checked this summary with co-father of the Internet Vint Cerf, he told me, 'Well, it is a *bit* more organised than that! But you are not far off!')

'This allowed people like me to live rich, full lives down at level one, while in parallel, at their own pace, people were making other developments' (Tim Berners-Lee's Worldwide Web was built up at the Application layer). In summary, the people who were developing web browsers didn't have to worry about the design of modems, and vice versa. Metcalfe's has similar vision for energy, which he's dubbed 'the Enernet', an energy system morphed to a more Internet-like model where 'energy will move from centralised power plants to distributed production'.

He believes it's the most sensible way to quickly upscale the amount of green energy we produce, while meeting ever-increasing demand. When Tim Berners-Lee gave the web to the world, he suddenly made it easy for more of us to access the Internet, and usage exploded. In fact, 'exploded' is too small a word: Internet usage went *supernova*, from one user of the Web (Tim) in 1990 to 3 billion people (or over 40% of the world's population) by 2014. With an Enernet in place, Metcalfe argues, we create an open system where green energy can flourish uncontrollably, allowing all those smaller, mostly renewable producers to get in on the game and play on a more or less even playing field, while nudging those dirty fossil fuels out of the picture, as seems to be happening in Germany.

'A lot of people talk about "going off-grid" but why would you want to do that?' says James as a couple more coffees arrive. 'For me that's like going off-the-Internet. The power of the network is having more people connected to it, not less. Dropping off is an extreme solution and it simply wouldn't work for lots of people who won't be able to generate the energy they need all year round, which would be most communities. So Open Utility is essentially a peer-to-peer marketplace for energy.'

'You're like the Airbnb of energy?' I ask.

'Yes, that's not a bad analogy. Airbnb aren't a hotel chain, Uber isn't a taxi company. We don't want to become a licensed supplier, we want to change the way the market works.'

James imagines a future where you might have a relationship with an energy company but they will act very much like your Internet service provider does now, simply giving you access to the network, then you select the energy you want to be delivered over it. It's still a world where big energy providers can play, offering their energy for sale like everyone else, but the difference is that as a consumer you can have a mix of suppliers, not one. This is why E.ON are busy buying renewable capacity like crazy. As the world shifts to a decarbonised world and renewables continue to fall in cost, they're seeking to position themselves as a cheap supplier in a newly liberalised marketplace. James isn't against the big players, but he's clear they'll have to adapt, or die.

Sound like a pipe dream? Actually it already works. The UK's first Enernet trial, in conjunction with Good Energy (the UKs leading 'green utility'), was a resounding success. A small selection of generators and business customers, mostly in the south of England, were given access to an open market, which they could interact with via Open Utility's Enernet software, Piclo. Customers could buy energy from the generators they chose (often local), while community generators were able to set their own tariffs (dealing with the biggest complaint I'd heard at Powering Up The North). The trial resulted in a partnership to offer an Enernet to all Good Energy's business customers.

James isn't the only one building an Enernet. In the Netherlands Vandebron launched a peer-to-peer energy marketplace in 2014 that, as I write, boasts seventy-five green energy producers providing energy to over 88,000 homes more cheaply than the big utilities. And there's more to come. In San Francisco, Geli is building an 'Energy Operating System'. In April 2015, Hangzhou

Zhongheng Electric of China announced it was raising $161 million, in part to fund an energy Internet cloud platform and energy Internet research institute. This followed Zhenya Liu, Chairman of the State Grid Corporation of China, outlining his plans for a 'Global Energy Internet' – essentially a peer-to-peer energy network for nations to trade energy, particularly renewables and specifically solar. Liu was clear – not only does he think the plan is feasible, even given 'transmission losses', it is a moral imperative in the face of climate change, and, of course, it plays well for a company that is buying up electricity grids across the world (and for the nation that manufactures the most solar panels).

I get the feeling that the Enernet is inevitable. Why? Because nothing boosts the economy like cheap energy, something Güssing amply proves. The reverse is also true. Most economists agree (which is rare) that sustained increases in energy prices signal recession as goods and services become more expensive. America remembers well the energy crisis of the early 1970s triggered by an Arab oil embargo in response to the US supplying Israel with weapons during the Yom Kippur War. Oil rose from $10 a barrel in 1970 to over $50 a year later. That crisis, coming on the back of a stock market crash, meant the US government was forced to employ oil rationing. Gas stations didn't serve customers on the weekends, a speed limit of 55 mph was introduced, some states even banned Christmas lights. The economy had been stiffed.

Governments and investors are beginning to see the big picture, and will increasingly back the idea of the Enernet and back it enthusiastically. There's a reason Open Utility received early funding to the tune of £500,000 from the UK's Department of Energy and Climate Change, and why Nicola Shaw, Executive Director of the UK National Grid told the BBC that 'We are at a moment of real change in the energy industry. From an historic perspective we created energy in big generating organisations that sent power to houses and their businesses. Now we are producing energy in those places – mostly with solar power.' The Confederation of British Industry's head of infrastructure, Michelle Hubert, agrees, seeing 'a profound shift towards a more flexible and dynamic system' with businesses and households becoming 'much more engaged in how they use, manage, and even produce energy'.

Corporate energy interests who try to resist the opening up of the market will find themselves suddenly on the wrong side of their own argument – no longer seen as powering growth, but throttling it. The bankruptcy of Kodak at the hands of the Internet and mobile phones didn't result in the world taking any fewer pictures, and the death of the energy players who fail to adapt to the Enernet won't see us with a lack of energy. As former Saudi Arabian oil minister Sheikh Zaki Yamani observed, 'The Stone Age did not end for lack of stone, and the Oil Age will end long before the world runs out of oil.'

He's right. Saudi Arabia's neighbour, the United Arab Emirates, will soon be generating solar energy at 3 cents per kilowatt hour, which is about half the current cost of coal and gas. In fact, the current Saudi government is in the process of restructuring its entire economy to live in a post-carbon future. 'Within twenty years, we will be an economy or state that doesn't depend mainly on oil,' says Deputy Crown Prince Mohammed bin Salman. Think about that.

So how does James feel, being at the cutting edge of a world-wide energy reboot?

'I have always been confident that this would happen and Open Utility would make it happen,' he says. 'Our confidence hasn't risen over time it's just other people's confidence in this idea has, to the stage where they think it's going to happen, too.'

'So, you feel good?'

His reply is a study in Herculean understatement. 'Well, I think it's interesting,' he says – and gets back on his bike to the office.

Bob Metcalfe makes the observation that, 'when we were building the Internet, in its early days, no one was talking about YouTube, nobody was talking about uploading family videos to be viewed by our friends'. There simply wasn't the bandwidth. A single video of a cat doing something funny would have taken hours to download. But, thanks to the architecture of the Internet, that layered model, it became possible to greatly increase that bandwidth to the point where we can now download movies in seconds, and have real-time video calls using services like Skype.

What happens when the energy system does the same – when we have an Enernet of cheap renewables, when we no longer have an Energy Trilemma? In short, it changes *everything*.

Nearly all of our economics and therefore most of our politics, are at their root the economics and politics of energy. What we pay for goods, our trading relationships, our squabbles over resources – all of them have a least one foot firmly rooted to the energy price. Create a world where energy is both cheap and no longer dominated by large corporate or national interests and you're looking at a very different world indeed.

'If the Internet is any guide, we're not going to be using less energy than we use today,' says Metcalfe. 'We're going to be using a *squanderable abundance* of it. What will be the YouTubes of energy?'

The first, Metcalfe suggests, is a worldwide version of the economic renaissance I've seen in Güssing. Another, he says, is 'solving global warming', because we'll have enough spare energy to remove CO_2 directly from the air. It sounds like science fiction, but it isn't. I know this because I sit on the advisory committee of the Virgin Earth Challenge, a \$25 million prize launched in 2007 by Richard Branson for whoever can demonstrate to the judges' satisfaction a commercially viable and environmentally sustainable way to remove greenhouse gases from the atmosphere. (Those judges include Branson himself, James Lovelock, Al Gore and Tim Flannery.) Until recently, the prevailing wisdom has been that removing CO_2 directly from the atmosphere would be impossible without using huge amounts of energy (making the idea prohibitively expensive). But, with a 'squanderable abundance' of renewable energy, coupled with the inventive approaches of the Earth Challenge finalists (at least one has been invested in by Bill Gates), who have reduced the energy needed massively, the game changes.

And what might we do with the carbon we take back? One option is to turn it into fuel. Again, it sounds almost unfathomable, but it's already been done. The car manufacturer Audi is making diesel from air and water at a pilot plant in Dresden, working with one of the Earth Challenge's finalists, Climeworks. It's using a version of the Fischer-Tropsch fuel synthesis process I saw yesterday, however, its source of carbon isn't vaporised biomass, but CO_2 reclaimed from the sky. Another approach is being pioneered by Chicago-based researchers Larry Curtiss and Amin Salehi-Khojin – a solar cell that efficiently turns CO_2 from the air directly into carbon-rich gas ready for subsequent conversion into liquid fuels. And – even more extraordinary – a recent experiment at the US Department of Energy's Oak Ridge National Laboratory accidentally came up with a way to convert CO_2 directly into ethanol using an electric current of just 1.2 volts with a conversion rate of 63%.

'Sky mining', as I call it, could become the ultimate recycling story, providing sustainable liquid fuels to accompany the shift to renewables enabled by a burgeoning Enernet. The likely timescale to market of such fuels, according to the Royal Society of Chemistry? Ten to fifteen years. (I check this estimate with Tim Fox and he thinks it's entirely reasonable.)

You thought the Internet was a big deal? Just wait for the Enernet.

I'm leaving Burgenland. The sun is shining, bouncing off the numerous solar panels I can see from my train window, and I'm thinking back to my meeting with ex-mayor Peter Vadasz.

'Is Arnold Schwarzenegger right?' I'd asked him. 'Should "the whole world become Güssing"?'

'I cannot speak for the whole world!' he'd said. 'But we can show another way. Perhaps a better way.' And then for a short moment his relaxed demeanour had lapsed, just a little, showing some of the fire that kept him and Reinhard pushing their 'crazy' idea for the best part of two decades.

'I was *determined* we wouldn't fail,' he'd said. 'Because, what then? Our opponents would write in newspapers, "Güssing has failed", and that would be a great shame. Because we've shown you can change the game, and that is something. This is another type of life. Today the idea of producing your own renewable energy is so normal here.' His smile returned. 'It's as normal as anything in the world, as normal as sitting here and having a beer'... and he'd raised his glass.

I'm happy, too. I've found another chunk of what I was looking for when I set out on my journey: a vision of a better future that's workable and touchable, a system-level innovation that's not a nice idea, but a day-to-day reality for the people I'm leaving behind.

At the heart of Güssing's revival is an issue everyone can get on board with. The town's transition was not prompted by a desire to reduce carbon emissions (although they have, by 90%), but to save its economy. Güssing, a town buffeted on the waves of geopolitics and left to slowly wither, is now largely in control of its own destiny. It occurs to me once more that the most remarkable thing about Güssing's energy system is how unremarkable it seems to everyone there.

10 ALWAYS BET ON THE TORTOISE

'You can cut all the flowers but you cannot keep spring from coming.'

– PABLO NERUDA, POET-DIPLOMAT

There's a kind of bar that you often see in gritty American TV thrillers – down-at-heel ale houses where criminals meet to broker drug deals and lonely figures sit mute, chugging one beer after another. Often they're basement establishments, adding a dose of claustrophobia into the scene.

The joint I'm in has a similar vibe, except the lack of natural light isn't because we're below ground, but because the windows are boarded up. From the outside I'd taken the place for a derelict, only realising it was open for business when my host Ashley ushered me inside.

Spying my MP3 recorder, she says. 'Just turn it off, put it away.'

'Because?' I don't get an answer.

'Why are the windows boarded up?' I ask as we take seats at the bar and order a couple of Dirty Blonde wheat ales.

She looks at me the way locals often do when an outsider has asked what, to them, is a naive question. By way of answer, she simply says, 'Welcome to Detroit.'

I'll admit it. I'm already beginning to wonder if coming here was such a good idea.

'You have to understand that, in Detroit, everyone's on the defence,' says Ashley. 'We assume everyone is out to get us, and act accordingly.'

'Act accordingly?'

'No windows in your bar, so it's harder to break in. Nothing nice or flashy on show, because people will take it.'

'Like my recorder?'

'It's not just that,' says Ashley. 'It's alienating if people think you're taping them. It won't play well here.'

'But I need to record …'

Ashley holds my gaze firmly and, in a tone that brooks no argument says, 'Just. Turn. It. Off.'

I look around the bar catching some glances from the locals … and put the recorder back in my pocket.

The next morning Ashley is picking me up from my lodgings, a room in Detroit's Corktown, the city's oldest surviving neighbourhood. Abigail the dog barks as I leave, happy it seems to see me depart. My hosts Cary and Jean-Marie tell me she's usually fine with their house guests, but a recent break-in has made the normally placid Labrador skittish and growly.

Burglary rates are nearly three times higher in Detroit than the US average, part of a wider crime problem that besets the metropolis. Assault is five and half times more likely, rape nearly triple. Murder rates are over eleven times the national mean. (If Detroit were a country, it would rank as third worst for murder in the world – only Venezuela and Honduras have higher homicide rates.) Taken as a whole the United States logs 38 crimes per square mile each year. In Detroit the figure is 389. Sky-high crime statistics are one of the indicators that tell the story of Detroit's broader problems, which become painfully visible in the morning sun. Driving west we see whole blocks standing empty; on others, a single still-occupied building remains. Occasionally small clusters of well-attended homes come into view, but these are few and far between and only serve to highlight the deterioration elsewhere. Detroit's population is 700,000, down from a peak of 1.8 million in the 1950s. The city's infrastructure is now simply too big for its dwindling inhabitants and is falling into disrepair around them. Bits of the city look like an ancient warzone, or a deserted rural border town, where weeds and vegetation run wild. It's eerie and shocking. The Detroit Blight Removal Task Force has recently recommended 22% of properties – vacant, damaged or dangerous – should be demolished. They're bulldozing the city. By the time I leave, I'll have a new perspective on these run-down buildings and vacant lots, but right now I'm overwhelmed by the decay.

How did America's biggest boom town become synonymous with urban rot and decline, subject of the biggest municipal bankruptcy in US history (with debts totalling roughly $19 billion)? In its heyday, Detroit was the world centre of automobile manufacturing, home to Henry Ford, General Motors and Chrysler, the fourth most populous city in the United States, testament to the power of American capitalism earning itself the sobriquet 'Motor Town' – from where another factory, pumping out classic songs instead of (now) classic cars, got its name: Motown Records, set up by former automobile worker Berry Gordy. But even Motown abandoned the city in 1972 for a new home in Los Angeles.

When Henry Ford and his contemporaries chose to base themselves here, the city boomed, attracting over a million new residents eager to find work, many of them black Americans migrating from the South, displaced by agricultural depression and keen to find a life free of mandated racial segregation (the former confederate states of the south had passed the racist 'Jim Crow' laws, disadvantaging blacks economically, socially and educationally). Their aspirations remained largely unfulfilled. Detroit became a cauldron of racial tension.

Historian Thomas Sugrue writes that blacks 'were closed out of nearly all white neighbourhoods' and the jobs offered to them tended to be 'the most menial, difficult, and dangerous'. Black communities congregated in 'older, deteriorating central neighbourhoods that had fallen out of fashion among whites'. Detroit became two cities – one black, one white – segregated both racially and economically. 'Blacks who attempted to cross the city's invisible racial boundaries regularly faced violence.'

As Detroit grew so did the city's membership of the Ku Klux Klan, with approximately 40,000 members in the city by the time Ford opened his mammoth River Rouge plant in 1928. The uprisings of 1943 of 1967 remain a deadly testament to the racial divide and contributed to Detroit's well-documented 'white flight', as swathes of the more affluent whites moved to the suburbs. This came in tandem with decentralisation of the automobile industry. Car manufacturers, wary of both racial unrest and a strongly unionised workforce, began to build factories elsewhere, while foreign competition and automation further reduced the city's labour market. Detroit slid headlong into economic and social decline, hollowed out by racism and the exodus of the industry it had built itself to serve.

'What's that?' I ask, spying one of the few multistorey buildings outside the business district.

'Michigan Central Station,' says Ashley.

It's an impressive piece of architecture. Eighteen storeys of French-inspired neoclassical design, created by the same people who imagined New York's iconic Grand Central Station. It's a huge monument to Detroit's former life as the biggest boom town in America. As we get closer, however, the splendour pales. It's derelict. The last train left in 1988.

Detroit has become a mecca for photographers of ruins and filmmakers looking for a backdrop that says 'urban apocalypse', something many Detroiters I speak to find distasteful. You don't have to have a PhD in empathy to realise that outsiders making money from Hollywood movies and coffee-table books trading on the fact your home looks (to them) like a war zone offers up a special kind of irritation, especially when you're trying to get the city back on its feet.

I learn very quickly that if you're not from Detroit you've got to prove yourself worthy, and quickly. Everyone tests me. Why am I in the city? When they find out I'm a writer, heckles rise. I'm here, no doubt, to write voyeuristically about the desolation, once I'm safely home in some swanky London pad. I understand their suspicion. The story of Detroit's decline has become something journalists of every hue seem to delight in, the literary equivalent of rubbernecking.

Ashley loves her city and the people in it with an almost messianic fervour. Their pain is her pain, their struggle is her struggle, their successes light her up. Her passion to improve the lot of her fellow Detroiters is what has brought me here, and being in her company certainly makes my passage easier. She's enormously well liked and respected. If Ashley says I'm OK, then I probably am, not that I get an especially easy ride from her, either.

'We vetted you,' she tells me.

I'd thought the long silences between my emails and calls while trying to arrange a visit were because Ashley was either too busy or (I feared) disinterested. Actually it was suspicion. She gets a lot of press enquiries. 'If you listened to my voicemail right now, there'd be a ton of them,' she says, and you can hear an undercurrent of cynicism in her voice.

'You're sceptical about the media?'

'Fifteen years of experience has taught me that most reporters are assholes – eighty per cent. Seriously.'

For years Detroit has been written off, the city equivalent of a failed state. But the truth is I'm not here to witness decay, but a story of seemingly implausible regeneration. Initially I thought I was coming to investigate another angle on how we might change our food and healthcare systems. Instead I'll learn the secret at the heart of nearly *all* successful system reboots.

Perhaps appropriately for a city dubbed 'Motor Town', the bulk of my morning is to be spent on a parking lot. This however, is unlike any car park I've ever visited. There are no spaces marked out, no barriers or ticket machines. In fact, you're not allowed to park your vehicle here, not that there's much space to, because in place of rows of parked cars there is, instead, a farm. Right in the heart of the city, in what should be the dominion of the resting automobile, I'm looking at two huge greenhouses, surrounded by nearly two acres of tomatoes, winter squash, melons, raised beds full of herbs, a cabbage patch and a commercial flower garden.

It's an incongruous setting for a display of such verdant abundance, especially for a boy like me who grew up in a small rural community surrounded by fields of crops and cattle. I'm used to seeing veggies growing in volume, but not slap-bang next to a freeway, in the shadow of a hotel complex on one side (plummeting real-estate prices have made Detroit a cheap venue for trade shows and corporate events), and a row of decaying office blocks on the other.

This is the Plum Street Garden and I've volunteered to do a morning shift harvesting some of its bounty, which has played well with Ashley. Over the next few hours, as we pick Sweet Basil, her demeanour softens and I ask her about growing up in Flint, a satellite city about 100 kilometres north of Detroit that suffered the same problems as its bigger neighbour. 'I was really struck by the effect of people leaving, how it changed the institutions, neighbourhoods, the landscape of the city,' she tells me. 'So as soon as I was done with college I went back to the neighbourhood I was raised in and started working as a community organiser.' Her work included creating and running the Flint Urban Gardening Land Use Corporation, helping local residents turn vacant lots into farms and gardens.

Swathes of the Greater Detroit Metropolitan Area (which includes Flint) are considered 'food deserts', defined by the US Department of Agriculture (USDA) as an area where at least 500 people and/or at least a third of the population live more than one mile from a supermarket or large grocery store. The USDA provides a handy online Food Desert Locator – a interactive nationwide map that Daphne Miller MD, author of *Farmacology* (a brilliant investigation into the relationship between our health and the way we grow our food), points out is 'almost identical to one showing zones with the lowest life expectancy and highest rates of obesity, diabetes and heart disease'.

In the definition of a 'food desert' the shortage is not, it turns out, calories. In fact, citizens who live in food deserts are generally the nation's fattest. While a depressing 71% of adult Americans are overweight (over half of whom are obese) the situation in Detroit is, almost unbelievably, much worse. In the poorest neighbourhoods, up to 90% of residents can be overweight, half dangerously so. It's another of Detroit's depressing roll call of statistics.

'If you want to find fresh fruit and vegetables in Detroit, you've got to travel much further than you would in a suburban community,' says Ashley. 'And when you get there you'll have less choice and have to pay more. That's a system problem.'

The US government's seemingly sensible response to this problem is to offer grants and tax breaks for grocery stores, small retailers, corner shops and farmers' markets. The flaw in this plan is that study after study shows that giving the poorest access to healthier food rarely translates into them buying it. As Barry Popkin, Professor of Nutrition at the University of North Carolina, points out, if you've grown up eating a certain type of food it becomes a habit. 'When we put supermarkets in poor neighbourhoods, people are buying the same food,' he says. 'They just get it cheaper.'

In fact, it's always been possible to eat well in Detroit. Local photographer Noah Stevens was inspired to create his photo-documentary *The People of Detroit*, showcasing healthy Detroiters of all classes, to make the point. Raised on welfare in the city, he watched poor diet 'cripple and eventually kill the people closest to me' and decided, as a result, to embrace a healthier lifestyle. The idea that a balanced diet is impossible here 'contradicts my entire life experience', he says. Noah, however, is rare in breaking free of the food culture he grew up in, as Detroit's and wider USA and global health statistics testify.

So, if despite the city's troubles it's always been possible to eat healthily here, and if increasing access to healthy food has little impact on whether people consume it, why, you might wonder, is there a farm on a downtown parking lot?

After two hours amongst the basil, we move to picking tomatoes, joining fellow urban farmer Willie Spivey. Willie's twinkling eyes, broad features and ready smile give him the sort of countenance that young children run towards in the hope of play and older folks raise a glass to. He tells me he's been working for just over a year at Keep Growing Detroit (KGD), the organisation that Ashley co-directs.

'How are you finding it?' I ask.

His answer comes like sunshine emerging from behind a cloud. First there's the smile and then, like warm tea flowing from the pot, he speaks. 'I love it. I love it! I tell you what – I love it so much I'm already looking forward to coming back to work tomorrow!'

Ashley, Willie and I all laugh.

'I never had a job like it! It took me till 60 years old but I don't call this work, I call it life! I feel good. It's a just beautiful thing.'

Born in the centre of Detroit ('Right in the 48206', he says, quoting the downtown zipcode of his birth), Willie spent most of his working life in the construction industry, but as the city declined, the work dried up. Today they're bulldozing the homes and office blocks he helped build.

'What did you do?' I ask.

'I was just walking around.'

'Just walking around?'

'Walking around. Simple as that. You know what I'm saying?' He pauses. 'Yeah. Drifting.'

'Because there wasn't any work?'

Willie fixes me with a stare. 'You're not going to get me to spit that red. Because there's work, you know, even if a person took a broom and a shovel from their house in the morning and chose to go out and make fifty dollars. You won't ever get me to say there's no work.'

It's an attitude you soon get familiar with if you meet urban Detroiters. The message is, 'Yes it's tough, but don't say there's nothing here. Don't tell me my town is empty, you don't know my town.'

Willie doesn't reveal how long he drifted for; he's much keener to talk about how he came to where he is now. In 2010 he retrained as an organic farmer, thanks to another (and slightly larger) urban farm in the city called Earthworks, 5 kilometres east in the Mount Elliot district. Once qualified, Ashley offered him a job with KGD to supplement the income from his own start-up farm, Uncle Willie Grows.

'So what's your job title?' I ask.

'I don't do titles!' he says. 'Titles lock you in. You got to stay humble, and sometimes titles don't allow you to do that. One morning God may say, "I have a different thing for you today".'

It takes me a while, but when I do find out Willie's main role at KGD I conclude that if there is a God they're probably pretty happy with him.

In 2013, Ashley was asked to speak at the annual TEDMED conference, the health-focused offshoot of the famous TED Global event. 'It was a horrible experience, *a horrible experience,*' she tells me as we drop our tomatoes into plastic pallets. 'I was listening to the guy before me and I was about to lose my lunch and tell them I couldn't do it. I was full of fear.'

Her nervousness came not only from the anxieties most of us have when asked to address a crowd (of which the TEDMED audience is definitely on the more intimidating end) but from a feeling she was being somehow traitorous to her community. 'I believe that the propagation of ideas within my community is far more weighty than an individual on a stage. I felt I couldn't do them justice and that by even agreeing to stand up there I was taking away from them.'

It's a tension I witness in Ashley over and over again. She wants me to understand that everything I've come to see is a collective effort, a shared struggle or victory, that there is no-one more important than anyone else. And yet her position as one of the founders of KGD, and her ability to clearly articulate that collaborative effort (and why it needs to be made), means she finds herself reluctantly in the spotlight. 'It's a double-edged sword,' she tells me wearily. I know that when she reads this chapter she'll hate being in it so much.

You'll never convince her, but Ashley's talk at the TEDMED conference was compelling stuff. She didn't need to be super-slick because she walked on stage with authenticity in bucket loads. 'We don't really practise what a lot of communities call allotment plot gardening, where *you* have your twenty feet, and *you* have your twenty feet,' she told the audience. 'The key here in Detroit is that people grow food *together*.'

Part of the reason Ashley was at a medical conference is because numerous studies prove that farmers (urban or otherwise) have a much healthier diet than their supermarket-sourcing neighbours. As Daphne Miller writes, urban farming 'accomplishes what grocery chains cannot: getting people to eat more fruits and vegetables'. For Western nations battling an increasing obesity epidemic, that's surely something to take notice of.

By some estimates, up to 10% of US healthcare spending is directly related to obesity. Increased body fat is strongly linked to a whole host of problems, including cancers of the breast, throat, kidneys, ovaries, pancreas and colon. If getting the populace more connected to the production of food is a potential way to stem the obesity tide, it's worth investigating. Given that a healthy diet is the most powerful preventative medicine in the world, Daphne found herself wondering why she had never prescribed gardening to any of her patients.

'Essentially Plum Street is a *bad-ass* teaching facility,' Ashley tells me. 'It has pretty high production for an urban farm, but it's also showcasing a lot of techniques that are important for growers to learn.'

The farm, it turns out, is part of a much wider network that trains Detroiters in all aspects of urban agriculture. Visit KGD's website and you'll find an almost overwhelming list of courses you can attend across the city, all part of the Garden Resource Program that's been running for over a decade. Courses here at Plum Street include education on irrigation and watering systems, how to cultivate strawberries and sweet potatoes, or learning from the wisdom of Willie Spivey, KGD's Head of Transplantation. Almost the entire length of one of the large greenhouses is given over to his seedlings.

'That's a lot of little plants,' I say, looking at the neatly tended rows of tiny leaves.

'Well, we support about 1,400 urban farms and gardens,' Ashley explains.

'So, Willie,' I say, 'your plants are across the entire city?' It turns out Willie is currently overseeing roughly a quarter of a million transplants.

'That's crazy, right? That's crazy but it's true, isn't it? But it's what I love. I love watching them seeds.'

'Willie will teach you how to put optimism into the soil,' chuckles Ashley.

'So, Willie,' I say. 'You're Detroit's Earth Mother? Is that it?'

He laughs. 'Hey now! Got to stay humble! What did I tell you about titles?'

Our shift finished, we take lunch at the Astro Coffee Bar, which sources its vegetables from Grown in Detroit, KGD's co-operative wholesale operation that aggregates produce from urban farms across the city, retails it and returns the revenue to the growers.

'So, KGD is about reconnecting people back to food, to improve their health?' I ask as I take a seat ready to eat my just-ordered potato salad. I figure it's only polite to try some of the locally grown produce.

'No, that's not what it's about, that's just one of the nice side effects.'

'But that's why you were talking at TEDMED, right?'

'Yes, and it's another reason I was uncomfortable. Health is not the reason we farm here. Our ambitions are much bigger than that.'

I take a forkful of salad and my mouth explodes. It's so damn fresh. 'Oh my God,' I say pointing my fork at the bowl.

'Good?' asks Ashley, smiling.

'I'll say.'

My experience, it turns out, isn't isolated. Inspired by the super-fresh produce being delivered by Detroit's urban farming movement, along with the availability of large empty buildings that can be cheaply repurposed as restaurants, the city is becoming a destination for gastronomes. The *New York Times* even went as far as to say that Detroit could no longer be considered a food desert, but was in fact, 'a culinary oasis' where keeping up with the dining scene 'is a full-time job'.

Lunch over and we're back in the car on our way to meet another key figure in Detroit's food renaissance that, I'm learning, isn't really about food. We take a roundabout route. Ashley wants to show me a couple of things.

We drive down Farnsworth Street, which immediately shows itself to be in better shape than a good deal of the rest of the city, a road of well-kept houses and mowed lawns. It's not an affluent neighbourhood, Ashley tells me, but it is, she says, an 'intentional' one – a street that decided to purposefully resist the wider decline, thanks largely to the catalytic influence of one man, Paul Weertz, a retired teacher (and urban farmer) who refused to yield, buying houses on the block and fixing them up rather than letting them rot. Drug dealers who moved into the area soon found themselves inconvenienced. 'Before I built a barn at the school, I didn't have a place to store the hay for animals,' he told the *Detroit Metro Times*. 'So I stored it in this crack house. I just filled the whole house with hay.' Now most of the community gets involved in renovating derelicts, which are rented to families who share the aspirations of their new neighbours.

'That's Rising Pheasant Farm, run by Jack and Carolyn,' says Ashley pointing to a couple of lots of tended greenery. Started in an attic, of all places, the farm is now a fully commercial operation growing kale, broccoli and lettuce, and a specialism in fast-turnaround 'micro-greens' (essentially baby herbs that pack a flavoursome punch), constantly in demand from the burgeoning restaurant scene.

After a few minutes we turn into a street that blows my mind. We pass a house covered porch-to-roof with stuffed animals. Another is festooned with large painted numbers and another covered in dots of all sizes (from the size of a ping pong ball to a truck tyre) and all colours. In a vacant lot a pink car is buried up to the door handles, while a tree, festooned with baubles, sprouts through the roof. Another lot hosts a myriad of freakish, but carefully placed bric-a-brac – shopping carts, children's toys, wheel rims, babies bottles, old radios. Two figures made from heavy-cast auto-parts stare impassively. Behind them a sign reads 'Oh the irony'. I feel like we're suddenly in a Terry Gilliam movie.

'This is the Heidelberg Project,' says Ashley.

'Yes, but what *is* it?' I ask.

'You decide.'

Part art installation, part protest, part urban rejuvenation, the Heidelberg Project is the brainchild of local artist Tyree Guyton who took a different route from Paul Weertz to reclaim the streets, turning the entire neighbourhood into 'an outdoor community art environment'. Guyton's dream is to 'transform the two-block area into a state-of-the-art Cultural Village', which isn't easy when nearly everyone is suspicious of you. The project's been partly bulldozed by the city government twice (although they're now on better terms) and suffered a string of arson attacks. Nothing is easy in Detroit.

We reach our destination: what appears to be a derelict gas station but for the odd façade – five panels of undulating waves of linked triangles, intermittently solid metal or open to the air. The frontage, I later find out, was designed by students at the University of Michigan 'to send a signal to the neighbourhood that there was something new going on'. The brief for the façade was apparently to develop 'something that can bring in light but that you can't throw a brick through.'

Walking inside I'm taken aback. I could be in hipster New York or London. There are rows of desks, each with a matching grey anglepoise lamp, at which are earnest twenty-somethings staring intently at their Apple laptops. Those not looking at their screens are scattered about in a small café area or in one of the internal meeting rooms with see-through walls that occupy a space previously reserved for trucks in need of love and attention. This is Practice Space, a start-up/co-operative hub where we've come to talk to Jess Daniel, the founder of Foodlab.

'Foodlab is the community of food businesses here in Detroit, but we're "after the farm gate" – our network doesn't grow anything,' she explains. 'Ashley works with growers, while we work with processors, distributors and retail. Essentially we're an incubator; we help them get started.'

Jess has a fascination for food systems, a passion she puts down to being half Singaporean ('Singaporeans are obsessed with food!'), 'dating a hippy in college' whose father ran a string of co-operative grocery stores and time working in Cambodia's Kampong district, where she discovered schools that put environmentally sustainable food cultivation at the centre of the curriculum. Returning to the US, she joined the National Sustainable Agriculture Coalition, which brought her into contact with Ashley. By this time she was breaking up with 'the hippy' and looking to settle somewhere for a while.

'The people I'd got to know in Detroit were saying, "You should come here", and I was thinking "That's *insane!*"'

'But you came anyway?'

'And it *was* insane!' she says laughing. 'First, I worked at somebody's farm, then I worked on a food truck that was illegal and I ...' She pauses as if she's not sure how much to share. 'Let's just say I did all kinds of weird things.' One of those 'weird things' eventually turned into Foodlab.

'A great way to meet people is gathering them together to eat, right? I like to cook, but the problem was I didn't have any money, so I needed people to pay for the food.' Her solution was to open an occasional 'pop-up' restaurant in her house, Neighbourhood Noodle. Jess's cooking is obviously good, because pretty soon she was hiring. 'We were going to open a restaurant proper, we bought a food cart … craziness!' The restaurant, however, got shelved, in favour of a bigger idea. 'I'd met all these people who wanted to start food-related businesses that were sustainable, ethical and community focused, so I just started helping. That's what I'm good at,' she says. 'Listening to what people want and then making that into something actionable.'

That's not an idle brag. Today Foodlab supports over 150 ethical food businesses created by local residents – from bakeries to catering services, coffee processors to distribution businesses, fisheries to food-based events companies, retailers and restaurants of all hues, along with a smorgasbord of food manufacturers, making everything from barbecue sauces to cheesecake. On the way here Ashley had pointed out Detroit Kitchen Connect, a Foodlab project providing shared commercial kitchen space for food start-ups. Foodlab also helps those fledgling enterprises apply for the numerous licences and certifications they need to start trading and offers development advice as they grow.

Ashley has already hinted that Keep Growing Detroit isn't really about growing food and, it turns out that Foodlab, at its heart, isn't really about selling it. Food is a catalyst; one way Ashley, Jess and everyone else I've met here are (they hope) helping Detroit build a bridge to a different future.

'So, you *have* to turn that off!'

My audio recorder is the subject of controversy again.

'Why?'

'Because, Mark' (Ashley emphasises my first name in a somewhat comic, teacherly fashion), 'you've got to live in the moment. That thing stops you engaging, and anyone else who sees it is going to hold back a little.'

'But I want to be accurate …'

'Just turn it off! You're missing the point. That thing keeps you observing, I want you to *experience* today.'

I realise that I either turn off the recorder or most of what it will do is document a rerun of this argument. Time to get old school. Notebook and pencil.

Early morning and I'm sitting in a ramshackle old house in Midtown, which now houses KGD's headquarters. Everywhere I look are boxes of seed packets. (KGD distributes over 60,000 of these packs each year.) On the floor of one room are three large buckets of garlic. Posted on one wall I find a map of the farms and gardens across the city, with dots in every neighbourhood. Outside a school bus is parked, ready to take gathering Detroiters, many exchanging horticultural tips, blueberry picking.

An hour later, about twenty-five of us are among the blueberry bushes chatting and bucketing sweet berries in glorious sunshine. This annual trip is one of the myriad food-related events KGD runs every year. 'We try to create big parties that inspire people to get to know each other and build relationships,' says Ashley.

Just as I'd found my relationship with Ashley begin to warm over picking sweet basil so the communication flows freely now – and these few hours in the sun will become one of my fondest memories of writing this book, including Ashley and I making sandwiches for everyone in a somewhat precarious fashion from a bumpy back seat of the bus. (Earlier we'd swung by Detroit's Eastern Food Market to pick up fresh bread and cheese along with some sweet peppery rocket bought from the Grown in Detroit stand.) It seems Ashley may have been right about turning off the recorder. I stopped observing relationships and started, if only briefly, to have them.

Our next stop is another of these communal events, a 'tomato festival' held at Detroit Farm and Garden, a locally run retailer of gardening and farming equipment, books and accessories. I'm momentarily shocked to see bottles of something called 'Beaver Nut Scrub', but this turns out to be a Michigan-made hand cream.

Local growers have brought a bewildering array of tomato varieties to the festival, from tiny berry-sized specimens to huge yellow examples the size of a fist. There are Green Zebra tomatoes, bright orange 'tangerine' tomatoes, tomatoes with names like Brandywire, Cherokee Purple, Early Girl and Great White. There's face painting, a wandering trumpeter, a tomato-tossing competition and later, I am told, a Tomato Queen will be crowned. The KGD stall boasts an impressive display of tomatoey produce, and tucked in one corner I notice with an irrational level of delight are the cherry tomatoes Ashley and I picked with Willie yesterday.

I don't get to see the Tomato Queen, as Ashley spirits me away to North End, formerly a nightlife hot spot, replete with speakeasies and after-hours gambling shops. Today the neighbourhood is dominated by empty buildings and vacant lots. Red's Jazz Shoe Shine Parlor (where legend has it the Miracles and Supremes used to sing acappella while waiting for a shine) still stands, but not much else. There's a makeshift stage on the vacant lot next to Red's, where a parade of musicians and poets ply their wares to a sparse but friendly crowd, part of a neighbourhood party. Ashley introduces me to Jerry Ann Hebron, who takes me on a tour of the next-door Oakland Avenue Community Garden. She tells me how the garden has become a hub for local projects, acting as a catalyst for a local farmers market, the renovation of a 'community house' for events and meetings, and an education programme based around sustainable agriculture and entrepreneurship. 'We want to reclaim the North End,' she tells me. I'm beginning to realise that pretty much wherever you are in Detroit there's an urban farming project or local food business no more than a block or two away – all part of the big ambitions Ashley and others have for the city.

Back on the stage a local poet is delivering fast and furious. I flip on my MP3 recorder, to capture some of it.

> 'When you ain't from my city and walk my streets then you will never know ... Elevated we stand fighting and bounce back even when we were down and out ... Allow me to reintroduce myself. I am Detroit, how's that for you?'

Back in the car I tell Ashley how impressed I'd been with Jerry Ann's story.

'Detroit has to heal from everything that's happened in the past. A person or a community can't do that unless they have the resources and a space to do so. We're trying to create those spaces through gardening, farming and food. Like I said, we don't do allotment gardening – the point is you come together.'

'So, it really isn't about the food?'

'No, but the food is good.'

'And it's not really about health?'

'No, but people get healthier.'

'It's about using a community-run food system as a way for the city to connect back to itself, to reclaim the story of the city from one of decay?'

'Finally, you're getting it!' says Ashley, but with good humour. 'A community that looks within for power is resilient – and connected.'

I'm hearing echoes of my recent trip to Güssing, where a

community that 'looked within' for power has, indeed, become more resilient and connected. There are also clear parallels with PatientsLikeMe, through which patient communities enjoy the same dividend of strength through kinship.

Sociologists would say that Ashley and her partners are trying to boost Detroit's 'collective efficacy' – that connected communities with common goals generally look out for each other and do better in terms of health, education and social justice than their divided cousins. Crime rates fall, people feel more connected, racial tensions ease. The dividends of increased collective efficacy are felt across every demographic. Richer neighbourhoods with less connected citizens suffer higher crime rates, for instance, than wealthy neighbourhoods with a sense of community. In *From Seeds to Stories: The Community Garden Storytelling Project of Flint* (a report on the Flint Urban Gardening and Land Use Corporation, the organisation Ashley had set up straight out of college), I find a quote from a local minister that sums up much of what I've seen here: 'This crossed every type of demographic boundary: race, age, and gender. People were becoming as entwined as the vines they were planting and, like those vines, were growing stronger by the day.'

'So, our goal is that Detroit becomes what we call a "food sovereign city" where the majority of the food we eat is grown locally,' says Ashley. 'We want the people of Detroit to be the dominant producer of the food they eat.'

It's one hell of ambition, but achievable. An analysis by Michigan State University's Center for Regional Food Systems concluded that by making use of the (nearly) 5,000 acres of vacant land in the city and maximising the growing season by, for instance, growing cold-tolerant vegetables like kale, spinach or carrots in winter greenhouses, it would be possible to grow 'roughly three-quarters of vegetables and nearly half of fruits' consumed by Detroiters. It's also worth remembering that Detroit is built on top of fertile soils, one of the key reasons French settlers stopped here in the first place. Some of the earliest maps of the area show more than fifty farms side by side, fanning out from what is today the city's waterfront.

When Güssing looked to itself for the seeds of its regeneration, it concluded its wealth of trees would be the genesis of the town's revival. In Detroit the abundant, currently under-used resource is land – and if the Detroit Blight Removal Task Force get their way, there could be a lot more available for farming. Another 2014 study, by the Detroit Food & Fitness Collaborative, suggests that boosting local production of food to 50% would keep an additional $409 million in the local economy. If manufacturing and processing (turning raw ingredients into cheesecake, for instance) could reach similar levels of local activity, the annual dividend would be more than double this, at $952 million, something Foodlab is working on. And of course, as I'd seen in Güssing, all that extra local revenue has a multiplier effect, generating jobs in nearly every other sector.

Detroit isn't the only city embracing the idea of urban farming. From Vancouver to Kathmandu, Cape Town to Hyderabad, Lima to Addis Ababa, urban farming is now being taken seriously as a way to rejuvenate communities and boost local economies.

Take Rosario in Santa Fe, Argentina. A city of 1.2 million souls, it shares many of Detroit's problems. At the beginning of the millennium Rosario was an industrial wasteland, with 60% of the population living in poverty. Residents of the city's slums were looting supermarkets for food. In 2002, the government launched an urban agriculture programme to encourage small-scale self-production of fresh food in low-income areas. Two years later 800 community gardens were producing food for 40,000 people, at which point the city established a rapid process for awarding grants to turn vacant urban land over to residents for agriculture. Ten thousand low-income families have become directly involved in gardening, and within a decade much of the city is reclaimed, all the while creating a new economy of locally grown food. A UN analysis of the programme concluded, 'there is widespread public appreciation of urban farmers as guardians of the land, whose work improves the living environment and contributes to the food and nutrition security of all citizens'. As an exercise in 'collective efficacy', it's hard to beat.

Outside my temporary Corktown home, Ashley and I say our goodbyes and I thank her for her time.

Her parting shot? 'Now remember when you write this up, try not to be an asshole.'

My next stop is Brazil, to look at what has to be the trickiest system in need of a rethink: government. But before I head to the airport I find a seat at the Brooklyn Street Local, its menu sourced as much as possible from the city's urban farms. The food tastes fantastic. I can't tell if that's because the ingredients are so fresh, the chef is a genius or because I feel more connected to the food I'm eating, heartened by its provenance and what it means for the neighbourhood I'm eating in. Likely it's all three.

I can't pretend to understand Detroit the way Ashley wishes I could, and I never will. As that North End poet had said, 'When you ain't from my city and walk my streets then you will never know ...' There's no getting away from it. Detroit still has a long way to go. (Proving just how hard it is to get new ideas off the ground, Practice Space, the trendy co-working hub where I'd met Foodlab's Jess Daniel, closes soon after I leave.) Wandering the city I'm still overwhelmed by the problems so obviously on view. Every other street corner seems to have a homeless person in residence. The ruined buildings that attract photographers of macabre architecture surround me.

But Detroit is far from down and out. It took Peter Vadasz and Reinhard Koch over twenty years to turn around the fortunes of the small town of Güssing by redefining its relationship with energy. In half that time, Keep Growing Detroit has grown from a network of eighty farms to one twenty times that size, and the momentum is growing. The task at hand is enormous, but the results are already showing. Who, for instance, would have predicted that a city long synonymous with urban blight would now be heralded as a food mecca by the *New York Times*? And I get the impression that Ashley and the communities here are determined to redefine the city on their own terms, however long it takes. At yesterday's tomato festival I'd asked Eitan Sussman, Ashley's co-director, about this and he'd said something seemingly enigmatic. Now I realise it's a perfect summary of her and, it turns out, many other systems innovators I've met on my travels: 'When it comes to innovation,' he'd said, 'always bet on the tortoise.'

11 HOW TO MAKE POLITICIANS POPULAR

'An imbalance between rich and poor is the oldest and most fatal ailment of all republics.'
– PLUTARCH, GREEK HISTORIAN

'Let me get this right,' I say, 'you robbed banks?'

I'll be frank, this isn't where I expected the conversation to go when I took my seat in Cantina do Lucas, a popular downtown haunt in the Brazilian city of Belo Horizonte for writers, artists and, in past times it seems, bank robbers.

'Yes, yes, of course,' replies Fernando, as if the idea of pulling a bank heist is no more unusual than popping out to the shops for a loaf of bread. 'We all did. We needed to raise money for the resistance. *She*', he says pointing and smiling, 'was a *terrorist!*'

The diminutive, softly spoken researcher sitting to my right, with her short-cropped, curly brown hair and rectangular thin-rimmed spectacles falls a long way short of the gun-toting, balaclava-wearing fanatic of popular imagination. In fact, Fernando is over-egging the dramatic pudding. Maria never took up arms, although members of her family did. 'It was a war. I was very young, maybe twenty,' Maria Inês Nahas says quietly. 'It changed me, changed everything. We all had to contribute, one way or another. My eldest brother, Jorge, took part in armed conflict and was in jail for two years. They tortured him.'

'This is Jorge who is joining us for dinner?' I ask. 'So he was released, he survived?'

'Yes, but it was a dark time. To get my brother and others out of jail, they kidnapped the German ambassador.'

The 'they' in question were Vanguarda Armada Revolucionária Palmares, one of the many active groups opposed to the Brazilian military dictatorship that ruled the nation between 1964 and 1985 – a regime that crushed political opposition via a combination of torture (including rape and castration), arbitrary arrest and imprisonment without trial.

Jorge arrives. In his mid-sixties and slim with a tidy white beard, this former guerrilla radiates an air of calm acceptance. Today he is part of the government, a retired surgeon who is both director of the hospital foundation for the Brazilian state of Minas Gerais (of which Belo Horizonte is the capital) and the city's Municipal Secretary for Social Policy. On his time as a 'terrorist' he's hard to draw out. It's not that he's deliberately reticent on the subject, more that it all seems such a long time ago, and he wishes to discuss his current preoccupations – namely government policy, the ins and outs of diplomacy, the mechanisms and frustrations of politics.

'Democracy is very complicated,' he says. 'It is disappointing sometimes, but at the same time ... well, it is at least democracy.'

'This is not a *real* democracy!' says Maria with fervour, banging her hand on the table. 'We *lost* the revolution!'

Maria's outburst seems odd, given that the Brazil I'm visiting holds free elections every four years with a voter turnout in excess of 80%. But Brazilians are far from happy with their democracy, believing it to be corrupt – and they're right.

On a scale of 0 (highly corrupt) to 100 (very clean), Transparency International, the global anti-corruption NGO, scores Brazil 38, highlighting significant problems with the independence of the judiciary, confidence in the rule of law, press freedom and ownership. In 2016, President Dilma Rousseff was impeached for allegedly manipulating government accounts to make them appear she'd met budget surplus targets set in Congress. There still rages a huge debate as to whether this is even illegal, and some argue the whole impeachment campaign is itself 'a parliamentary coup brought about by corrupt politicians'. This comes on the back of the largest corruption scandal in the country's history, with politicians and directors of the petrochemical behemoth Petrobras (itself 51% government owned) accused of colluding on suspicious transactions totalling $22 billion. The accusations involve kickbacks taken from construction companies in return for contracts, with some of that money finding its way to 'friends' in the governing coalition. More than thirty members of Congress are implicated.

Brazil's woes are part of a wider, global crisis. The Economist Intelligence Unit's (EIU) Democracy Index, reminds us that only 12% of the world's population live in 'full' democracies (a number that's been declining in recent years). A further 39.5% of us are governed by 'flawed' democracies with significant weaknesses in governance (of which Brazil is one). 17.5% of the world's people must suffer 'hybrid regimes' like Pakistan, where free speech is curtailed and the judiciary are unduly influenced by the powers that be. At the bottom of the pile are 'authoritarian regimes' where 'political pluralism' (the peaceful coexistence of differing convictions and lifestyles) is actively discouraged, civil liberties abused, the media controlled by the ruling regime, criticism of the government punished and censorship commonplace. (By these criteria Russia remains an authoritarian state, along with 50 other nations.)

I've come to Brazil to investigate an innovation many believe could be one solution to democracy's woes. Personally I'm sceptical. It's not just the backdrop of corruption. 'Participatory budgeting' (PB) as it's called, looks dull as dishwater, involving arguably the twins pillars of extreme boredom – spreadsheets and committee meetings. On top of all that, it's claimed it does something most people would consider impossible.

It makes politicians popular.

Philosopher Roger Scruton writes, 'if we study the words of Western politicians, we will constantly find that the three ideas – democracy, freedom and human rights – are spoken of in one breath, and assumed in all circumstances to coincide'. He points out, however, that it is only under certain conditions that they actually do so. Those conditions include a justice system free of political interference, the legal protection of basic civil liberties (including freedom of speech, movement and religious worship), working property rights (allowing individuals and organisations to legally own and earn income from both physical and intellectual assets) and, via a free and fair electoral process, a legitimate opposition that can challenge the existing administration and replace it if the people wish.

But even nations that have attained this democratic nirvana fail their citizens horribly. Take the inability (or unwillingness) of elected representatives to halt the increasing divide between the very rich and everyone else. Today the most affluent 10% of the US population control roughly 75% of its wealth (with the most well-off 3% representing over two-thirds of *that*) – leading some analysts to dub the USA the 'Unequal States of America'. In the UK the richest 10% of households account for 45% of total aggregate household wealth, while the bottom half are left to share just 9%. And the celebrated democracies of Norway, Sweden and Denmark are no better, with their top 10% holding between 65% and 69%. In fact, there's a fast-widening gap between the rich and the rest pretty much the world over, whether you live in a democracy or not. The World Bank's GINI Index (a popular inequality measure named after its inventor sociologist, Corrado Gini) tells us when it comes to wealth distribution the USA's performance is roughly on a par with the corruptly governed Republic of Congo and the UK's with the repressive administration of Burundi.

Another stratospheric failure of modern democracies has been their inability to provide a financial system that serves the majority. One has to wonder how the US Department of Justice has been far more successful in pursuing wrongdoing at FIFA (a footballing organisation based in Zürich) than it has on Wall Street. So far, just one financial executive in the USA has been jailed for their role in undermining the global economy as part of the 2008 global crash which destroyed the livelihoods of millions. With a track record like that is it any surprise barely a quarter of US citizens really trust one of those crucial pillars of democracy: the courts and the criminal justice system?

What about those other institutions that underpin democracy? I'm citing statistics from the 'Land of the Free' not only because it likes to see itself as the world's most valiant defender of democracy but also because problems with governance in the USA are echoed in many other democracies. For example, its difficulties with the press. Once diversely owned, the US media is now almost entirely controlled by six companies – all with their own partisan take. Citizens are justifiably wary, with 74% regarding newspapers with caution or outright cynicism (a figure that rises to 78% for television news). What about that other foundation of a working democracy – functioning property rights? Well, as mentioned above, property rights (whether they're physical or intellectual) do work *very* well ... for the super-rich, who have snaffled up the majority of them. In this context it becomes easier to see why Britain voted to leave the EU and how Donald Trump was able to become US president. For large numbers of the population, democracy isn't working and anyone who offers to give it a good poke looks a better bet than voting for more of the same.

In his famous Gettysburg Address during the US Civil War, Abraham Lincoln called upon those listening to resolve that the war dead should not have perished in vain, and that 'this nation, under God, shall have a new birth of freedom, and that government of the people, by the people, for the people, shall not perish from the earth.' The poor man is probably turning in his grave. Increasingly, modern democracies (the USA included) cannot be argued to be 'government of the people, by the people, for the people' but rather 'government of the people, by very few disconnected people, for the disproportionate benefit of the rich', explaining perhaps why the number of US citizens who have a generally positive view of Congress is just 9% (seriously).

Citizens of democratic countries, while often happy with the *concept* of democracy, are far less enamoured with its implementation. They trust their parliaments less than the constitutions those parliaments are supposed to uphold, their sitting governments less than their parliaments, and their political parties less than their governments. Sweden, regularly held up as one of democracy's stars, is a good example: 81% of Swedes are 'satisfied with democracy' but only 32% 'trust political parties'. You could perhaps summarise the problem by saying, 'Most people think democracy would work fine, if it wasn't for the politicians.' Worse, many younger citizens are now cynical about the very concept of democracy: when polled, only 30% of US millennials rated living in a democracy as 'essential'. And so voter turnout is dropping worldwide. There's a direct correlation between the level of satisfaction citizens feel about the way their democracy works and their propensity to vote, creating a vicious circle that undermines the whole enterprise further.

In my experience, many inside the political class confuse disenchantment with indifference, 'voter apathy' being an easier pill to swallow than the admission that our democracies are increasingly unfit for purpose. If the democracy you live in can't save you from the injustices of inequality, if its regulators are lame ducks, if it can't find ways to punish those who criminally mismanage the financial system, and if the media (supposedly one of the institutions that holds everyone to account) peddles partisan untruths at you with depressing regularity – and this all happens *no matter who is in power* – why vote for the usual suspects, or at all?

I'm staring at a lightbulb. There's nothing special about it, except for its location – about twenty feet above us and suspended on a wire strung between two wooden poles on either side of a narrow dirt road. Looking down the street I see several more domestic bulbs of varying designs hanging off a cobbled-together system of wires.

'*Gato*,' explains Duval.

I've enjoyed getting to know Duval Guimaraes. Born to ranchland farmers in the western state of Mato Grosso, he moved to California to study Political Science and Liberal Arts at Santa Barbara City College, transferred to the American University in Washington DC to study International and Government Relations, went on to the World Bank, then the IMF, before completing a masters in public policy at the Harvard Kennedy School of Government. Now returned to his native Brazil and barely 30, he's special advisor to the mayor of Belo Horizonte. He's also acting as my translator for the day, so I feel I've rather lucked out. But I suspect he's having a joke with me. My Portuguese is (very) limited, but I do know that 'gato' means 'cat', not 'lightbulb'.

'It's also Brazilian slang for pirating public services,' Duval explains. 'It's very common in the favelas. They tap their wires into an official outlet and draw the electricity off. Often it's streetlamps that act as the source, but there aren't even those here, so they've made their own.' It's not only electricity they hijack. I spot networks of jerry-rigged white piping and plastic barrels that turn out to be a neighbourhood-wide 'gato' water system.

I'm in the Capitão Eduardo district, home to many of Belo Horizonte's favelas and 'unofficial' neighbourhoods, as well as being, Duval tells me, 'one of the most dangerous areas in the city'. That's not reassuring given that Belo Horizonte is ranked as one of the fifty most treacherous cities in the world, with a murder rate over twenty times higher than my native London. The biggest contributing factor to that statistic is the kind of poverty I'm seeing around me, although this is apparently one of the 'better' examples of the numerous slums in the city.

Earlier in the day at City Hall, Rodrigo de Oliveira Perpetuo, the city's Municipal Secretary for International Relations, had told me how Belo Horizonte was the first planned city in Brazil and 'a symbol of the new republic that was emerging in the nineteenth century'. Originally built to host 200,000 people, today the slums alone account for over one and a half times that number while the total population pushes 2.5 million residents.

To enter the neighbourhood we'd had to navigate the only vehicle entrance, an incredibly narrow dirt passage that our car made it through with barely an inch to space, and clearly too narrow for a larger vehicle like a bus, police van or ambulance. 'That is one of the first things we would change if we could get PB,' a local resident tells us. 'If there's an emergency you can't get in or out. I'm going to have to buy a helicopter!' she jokes.

A wiry lady in a dress printed with large tropical flowers on an immaculate white background, she stands out against the unpainted brickwork and brown clay of what passes for streets here. It turns out she's president of the local Community Association, a community she tells us doesn't officially exist. All the property rights here are informal, with no standing in law. Legally speaking, the favela's buildings and the 4,000 people in them are invisible.

'Look!' she says, pointing at the *gato*, 'We have to steal our energy and water because no utility will work here until the city passes a decree to formalise us. The electricity has no stability, it burns out our equipment. We have no official address, we cannot prove we live here. You cannot call the police here. To register my daughter at school I had to use the school's address.' What she says next you may think is a joke, but her face is deadly serious. 'Let me tell you. I'll pay my first electricity bill very happily. I'll put the first electricity and water bills in a frame and hang them on the wall.'

Duval turns to me. 'It's such a different reality, man. It puts things in perspective ... all the things we take for granted. Sometimes I wonder if the mayor really gets it.'

Despite Duval's barely concealed cynicism about his boss, Belo Horizonte has long been seen as a beacon of progressive policy when it comes to improving the lot of the urban poor, and former revolutionary Maria Nahas has been instrumental in helping local politicians 'get it' thanks in part to her training as an ecologist. That morning on our walk to City Hall I'd said to her, 'I'm guessing becoming an ecologist helped you think in systems? Every organism is a system, but you were also interested in how each organism fitted into a bigger system?'

'Yes!' she'd exclaimed, as if this thought had only just occurred to her. 'You are absolutely right! In fact, I was an excellent taxonomist! I saw everything as a system.'

Taxonomy, the discipline of naming, classifying and describing organisms, and thus identifying their relationships within a wider network of life, is a key foundational skill in ecology, but for her doctorate Maria approached it with a unique slant.

To say Maria is obsessed with statistics is an understatement. To be more accurate, Maria is consumed both with statistics and the *methods* by which we collect them. For the last twenty years, starting with her doctoral research, she's been developing ways to name, classify and describe the conditions of Belo Horizonte's citizens. In short, she's created a way to taxonomise inequality.

Today she is most famous (in urban development circles) as the creator of the Índice de Qualidade de Vida Urbana or IQVU (Quality of Urban Life Index). I'd encountered a mini-IQVU in Detroit, the US DOA 'food desert' map. Maria's index includes similar data, detailing citizens' access to grocery stores, supermarkets and food outlets across the city, but that's a small subset of the information she's been collecting. Covering the entire city, there are scores for how easy it is to access (and the quality of) education, housing, sanitation, electricity, communications, healthcare, financial services and culture, along with indicators of levels of crime (of various categories), traffic safety and noise pollution. This provides the city's policymakers with a regularly updated picture of the impact (or not) of their investments.

Working with teams across the city, Belo Horizonte's IQVU was first calculated in 1994 and has been continually refined ever since. It's regularly cited as best practice, with Maria called upon by various United Nations bodies and international universities to share her expertise. Her work found an early advocate in the former mayor, Fernando Pimentel. Another veteran of the resistance linked to the guerrilla group VAR-Palmares, he was also imprisoned and tortured. Upon gaining his freedom he studied Economics before rising through the ranks of city government and enjoyed widespread popularity (becoming hailed as one of the best ten mayors in the world), due in part to his enthusiasm for embracing Maria's findings and so actively directing the city resources to its most needy.

'The IQVU allows us to see regions of the city where the supply and access to services are poorest, identify which of those should therefore be prioritised for investment, and *what* to invest in,' says Maria with more than a hint of pride in her voice. So far, so very ordinary you might think. Surely using this kind of statistical analysis to guide spending is commonplace? You'd think, wouldn't you?

'Which government wants to be transparent?' asks Maria with a shrug. 'Most won't want anything like an IQVU. They want to spend money as they want.'

It's true that governments *do* collect a lot of data, but those elected are often far from keen to use it for the simple reason that honest hard facts can reveal truths that are awkwardly at odds with the prevailing ideology of the party in power. Another reason some governments don't like facts, especially if they're in the public domain, is because they make the whole business of corruption so much trickier and, as Transparency International's Global Corruption Barometer reveals, six billion of us live in a country 'with a serious corruption problem'.

So, what's the solution? A popular analysis shared by many economists, social scientists and NGOs (including Transparency International) is that, since our failing democracies disproportionally visit their worst failures on the poorest, it is only by involving the most disadvantaged in the business of government that the problem can be effectively addressed. I've come to Belo Horizonte to see how Maria and her colleagues have done just that.

It all started more than thirty years ago. In October 1985 the first free elections since the end of military rule earlier that year were on the horizon. Mayoral candidates for the city of Porto Alegre, the capital of the southernmost and historically rebellious Brazilian state of Rio Grande do Sul, were meeting community activists, and in particular the União das Associações de Moradores (Union of Neighbourhood Associations), who had a radical suggestion for doing things differently.

'They must have had little idea of how the discussion that evening would resonate around the world over the next quarter century', write sociologists Ernesto Ganuza and Gianpaolo Baiocchi. The activists came with questions like, 'How would the candidate, as Mayor, improve public housing ...?' which, given 'a third of the city's population lived in slums' and 'few had consistent access to clean water', were very reasonable questions to ask. But there was one question in particular that would eventually lead to the city earning its reputation as a beacon for democratic reform.

Earlier in the year at their annual congress, echoing the modern belief that one of the key ways to fix democracy is to involve the poorest in its operation, the Neighbourhood Associations had agreed and endorsed the collective notion that there must be some level of community control over municipal finances. How, they wondered, would the new mayor implement this?

There was no immediate answer. The first Mayor, Alceu Collares, fudged the issue and was ousted two years later in favour of Olívio Dutra, whose campaign had promised the creation of 'popular councils' made up of ordinary citizens. After a shaky start, Dutra's administration oversaw a series of public assemblies across sixteen districts. Citizens were asked to put forward their ideas for public works and also elect one person out of ten to sit on a municipal budget forum, which would consider the detail of implementing them. The result was the Public Works Projects Plan (Plano de Obras), not only detailing which investments should be made, but also acting as a document to hold the city government to account.

As prosaic and far from perfect as the process was, the fact it had been made real, supporting the original demand for a degree of community control over municipal finances, was a radical step. In a city beset by poverty, citizens (many from the most disadvantaged constituencies) had become deeply involved in deciding how to spend public money, funnelling it into projects that would have a direct impact on their day-to-day reality – with the ongoing ambition that every year the people of the city would be invited to co-evolve a budget for new public works and the process would remain open to *anyone* who wanted to get involved. There was very little precedent anywhere else.

Adventures in 'direct democracy', where mechanisms exist for ordinary citizens to decide on matters of policy and spending, are few and far between. The most notable is over 2,000 years old, Athens' Citizens' Assembly, or Ekklesia. The Ekklesia met roughly once every ten days on the Pnyx (a hill above the marketplace) to decide on everything from judicial to financial matters. It was the state's ultimate governing body, a government made of citizens. It's important to remember who qualified as a 'citizen', though – not women or slaves for starters.

Democracy then moved west to Rome, morphing into a form we're more familiar with today – a 'representative' model where we vote for parliamentarians who are entrusted to decide things for us on our behalf. Roman citizens voted for 'consuls', who were then allowed to elect Senators – almost all of whom were chosen from the ranks of the wealthy or the aristocracy and who got to stay in power until they died. So if you were a man (again, women and slaves weren't considered citizens), you could vote for someone (a consul) you wanted to take part in a *further* election (which you weren't party to) that almost invariably put a wealthy toff into office.

Two and a half thousand years later a slightly more benign version of this representative system, mediated not by consuls but by political parties, is the prevalent model of democracy throughout the world ... and almost invariably puts a wealthy toff into office. As Olívio Dutra remarked, the current system 'reinforces the idea of occasional and incidental citizenship, restricted to the act of voting and that the elected, instead of representing, replaces the voter.'*

Porto Alegre's process of Participatory Budgeting (PB) seemed to offer an intriguing halfway house, providing a workable interface between those elected to government and the citizenry, and it soon gained interest around the country. Other city governments took notice, hungry for a democratic revolution

---

* Today the Swiss political system is the nearest thing to a national direct democracy operating anywhere in the world. It's still representative (laws are made in Parliament by elected officials), but citizens are entitled, if 50,000 of them offer their signatures, to put almost every law decided by their representatives to a general vote. They can also demand a change of the constitution if 100,000 signatures are gathered.

that would act as the strongest counterpoint to the recent memory of military rule. Had a city blighted by poverty found one way to do 'government of the people, by the people and for the people'? Perhaps. There was just one snag. Porto Alegre's participatory budget was an unmitigated disaster.

On taking office, Dutra inherited a calamitous set of public finances thanks to a spectacularly underhand move by his predecessor. The outgoing mayor, Alceu Collares, had given city workers an enormous raise following his defeat in the election, but before he left office, leaving Dutra 'with only 2% of the city budget available for badly needed public works projects'. With so little to spend, the stage was set for inevitable disappointment. Almost nothing got done. The Plano de Obras was little more than a work of optimistic fiction. Faith in participation and Dutra's administration quickly faded. The next round of citizen assemblies, now seen as little more than an exercise in lip service, saw attendance evaporate. Porto Alegre's brave experiment in direct democracy was all but dead.

Dutra recalls how his office justified the government's difficulties by explaining that 'the municipal budget was like a short blanket; if pulled up it would uncover the feet, if pulled down it would uncover the head'. Upon hearing this, a local textile worker responded:

> 'Of making blankets I understand a little. At the factory we know the width, length and thickness of each blanket that needs to be done. But that blanket you are speaking about never passed through our hands. I suspect that if we could help, it would come out better.'

Demonstrations in front of city hall called for the Plano de Obras to be honoured and the Head of Public Works to resign. The people's fury was palpable, but with so little money in the coffers there was little Dutra could do. His preferred solution to the city's financial woes – a more progressive local tax system that would deliver much-needed extra funds into the city's accounts – was probably going to be impossible to implement thanks to strong opposition from parties on the city council who favoured tax cuts they believed more palatable to the electorate.

On the day the council gathered to vote on Dutra's tax proposals the protesters were out again, but this time to *support* the mayor. Local historian Humberto Andreatta writes, 'a crowd gathered in front of the City Councillors Headquarters, demanding the executive's proposal be kept. This event caused an opposing city councillor to remark: "This is the first time I have seen the population struggling for a tax *increase*".' The people had come to help Dutra with his 'short blanket' problem and miraculously he got his tax overhaul, and a 50% increase in tax revenues as a result. Around the same time he negotiated a more generous settlement from federal government.

Slowly, the assemblies began to fill up again. Citizen-elected 'budget delegates' held their own regular district forums where local people could take more time to discuss and prioritise suggested investments. A simple set of criteria (a proto-IQVU) for directing more money to the most needy areas and addressing the most pressing concerns was also evolved. The result was a complete turnaround. Pretty much everything on the 1991 Plano de Obras got built, and the mayor's popularity soared. Dutra and his team had pulled a rabbit out of a hat. More importantly they'd lit a touch paper.

I'm in the upstairs room of a busy health centre in Belo Horizonte's Ribeirio de Abreu district, a poor community of

9,000 just down the road from Capitão Eduardo. I'm meeting two community 'budget delegates', Eduard and (another) Maria, who represent different neighbourhoods in Belo Horizonte's participatory budget. Eduard is trying to get a word in, but it's not easy. Maria, burly and ebullient, is clearly used to dominating. Duval is doing his best to translate.

'Political paving!' she exclaims. 'That how it used to be! Around election time you might get a bit of paving done, but without any accompanying sanitation work. It looked nice but it wasn't what we needed. For those people in the margins, where the investment was needed the most, there was *nothing!*' She points a finger accusingly, though not at anyone in particular. 'There were lots of investments in the centre of the city, though! – where the planners work!' This is followed by a contemptuous snort.

'So how is it now?' I ask. The answer comes at me like water from a fire hose. Maria wants to tell me everything about Belo Horizonte's participatory budget, including everything that's wrong with it (which, according to her, is a lot), how she'd like more say in every aspect of the process, and how taking part isn't easy. 'Exercising your citizenship is very hard. Only a few of us are willing to make it happen.'

Despite her complaints, Belo Horizonte's participatory budget, the second to get going in Brazil, has over the last two and a half decades evolved into a modern-day adaptation of Athenian-style democracy, minus the sexism and slavery, and with added statistics and bus trips. If Porto Alegre is the spiritual home of participatory budgeting, then Belo Horizonte is its biggest church, lauded the world over as an exemplar of best practice, and I've come here to find out how it works on the ground.

Between them, Eduardo and Maria explain. Every two years the city calculates the funds it has available for new works, allocating them according to Maria Ines Nahas' statistical powerhouse and reality checker, the IQVU – ensuring those areas with the lowest IQVU scores have the most money put on the table. Neighbourhood assemblies are called where citizens can put forward their ideas for new investments. A further regional meeting (representing groups of neighbourhoods) whittles these long lists down to a total of twenty-five projects per region, out of which fourteen will be chosen to receive funding (112 in total). 'So, neighbourhoods compete!' exclaims Maria. 'There is never enough money for everything. That can get ugly.' The city helps cost each proposal, and then democracy gets on a bus.

'Over a couple of Sundays the city provides buses for all the elected budget delegates to visit the proposed projects,' says Eduard, finally getting a word in. 'We visit all twenty-five in our region and the people proposing them tell us why they think their idea needs to get done.'

'We give up our weekends and evenings a lot!' interjects Maria. 'But you do it because you believe in it!'

'It's when you get on the bus, you begin to build empathy between neighbourhoods,' Eduard explains. 'You can see they need their investments, and sometimes they need them more than you need your's, so at the next regional forum some people remove their own proposals. It's not always pretty. People who lose can leave shouting and crying and cursing.'

'I remember losing and saying I would never participate again,' admits Maria, 'but I came back. It is democracy.'

'Let me tell the truth,' says Eduard. 'The year the budget funded this health centre I was advocating for another investment, a street improvement project, but I gave it up to focus on getting this centre. This was needed more.'

The bus trips (called locally 'the caravan of priorities') are one of the cornerstones of Belo Horizonte's participatory budget –

and I'm heartened to hear about them from the horse's mouth. This is politics that isn't dominated by 'left versus right' debates. Instead, people who live in the neighbourhoods where money is to be spent agree who needs what the most and change their positions accordingly. It generates less insular thinking.

Will Eduard be trying to get his street improvements back on the agenda next year, I wonder? He frowns. 'Some projects take years to happen. It's too long.' He taps his moustache thoughtfully. 'No, I've hung up my boots.'

Maria snorts again. 'He always says that! And each year he's back!'

Perhaps that's because the results are plain to see, including the building we're sitting in, which, its manager tells me, has enough capacity to meet the neighbourhood's needs.

'You won't be asking the participatory budget for another one, then?' I ask.

'No, this is plenty,' she says. 'Really, it is. It's well used and that's good. We had nothing before.' She pauses. 'But an outdoor gym would be nice …'

Since 1993 Belo Horizonte has allocated over $600 million via participatory budgeting, with the resulting work overseen by a network of citizen committees called Comforças. Currently, roughly $80 million of the city's budget is determined by the process every two years. The city has also experimented with a separate participatory budget to specifically address housing issues, as well as instigating a famous (if you're an urban development and democracy geek) online participatory budget to consider citywide investments. More an online referendum than full budgeting process, this ran in 2006, and again in 2008 attracting nearly 10% of the electorate. Again, the city was keen the poorest voices were heard, funding a bus filled with Internet-connected computers to visit deprived areas and encouraging people to cast their vote on proposed projects. There's also a Children and Adolescents Budget being trialled in sixteen schools across the city.

The main attraction, however, remains the regional participatory budget, involving up to 40,000 citizens with high levels of participation in the most disadvantaged groups. 'Today 80% of the city's population are living within half a kilometre of infrastructure financed by the participatory budget.' Indeed, I spend the rest of my time here visiting projects funded by the process, often on streets paved by it, and never travelling more than a few minutes between them. A new community centre with a natty basketball court, a new school, a much-needed bridge to help citizens get to work quicker ... I can't get round them all. The various participatory budgets have delivered well over 1,500 projects so far, from entire housing developments for the poor, to simple footpaths, from sports complexes to public squares. In my hotel room that evening, I download eighteen years' worth of IQVU statistics. Pretty much every neighbourhood shows an improvement. Almost across the board, violence is down, while access to public services, sanitation and culture is up. It's hopeful reading.

Despite my initial scepticism, I have to admit that I'm rather taken with what I've seen of participatory budgeting here – citizens spending their own taxes to shape their own neighbourhoods, areas in which they are the real experts. As the White House's Open Government and Innovation advisor Hollie Russon Gilman writes, it 'recasts politics on a more human scale'. Even so, it's important to ask if Belo Horizonte or Porto Alegre are better places to live in 2016 than they would have been without participatory budgeting. Advocates of the process point to the fact that after eight years of participatory budgeting in Porto Alegre half of the city's unpaved streets had been seen to, students in elementary and secondary schools had doubled, public housing construction had accelerated and bus companies were now serving previously neglected communities. Another success: today 94% of households in Porto Alegre have access to adequate sanitation (up from 49% in 1989, when those neighbourhood activists first proposed the idea of a people's budget).

But might the investments made via a participatory process have been made anyway? With Maria's IQVU to hand, could Belo Horizonte have done away with all that pesky (and expensive) consultation? As Josh Lerner, author of *Everyone Counts – Could Participatory Budgeting Change Democracy?*, points out, 'it is infinitely easier for a handful of staff to set a budget than to engage thousands of people over a year-long process'. Do regular citizens *really* make better choices than city governments? We've all indulged in the fantasy of what we'd do if we were in government, and how much better things would be if we were, but, as Winston Churchill famously said, 'The best argument against democracy is a five-minute conversation with the average voter.' And if you're single, and wish to remain so, do check out Kenneth Arrow's 'impossibility theorem' and Philip Pettit's 'discursive dilemma'. Influential theories in the realm of political science, taken together they argue that there is no voting system that can accurately reflect society's views, and, even if there was, those views would be wrong because the people voting can't ever know enough to do so in a sensible way.

It's an oft-stated view in political science circles that most citizens simply have no interest in being politically engaged. While Belo Horizonte achieves a participation rate for the PB of 40,000, it's worth remembering the city's population is 2.4 million, which means roughly 1.7% take part. Of those, only one in ten turn up after the initial assemblies. The individuals involved in the real detail is a measly 0.17% of citizens. A similar proportion get involved in Porto Alegre. So is participatory budgeting no more than an expensive way to help inexpert neighbourhood busybodies feel important?

The first thing to clear up is the question of whether participatory budgeting results in better outcomes for citizens. The good news is there's plenty of evidence to look at. In the thirty years since Porto Alegre started the ball rolling, participatory budgeting has spread around the world. By 2000, 130 cities in Brazil had adopted some form of the process and, as the new century got going, participatory budgeting's international footprint grew rapidly. Notable adopters include Rosario in Argentina (also, as I discovered during my time in Detroit, an urban farming pioneer), where since 2003 nearly 90,000 citizens have taken part in deciding how to spend $9 million each year. In Seville residents decide on approximately 50% of local spending (roughly $19 million a year) for their city districts. In December 2013 the White House identified participatory budgeting as best practice, including it as a cornerstone of the Open Government National Action Plan. Paris has launched its Budget Participatif, in which 40,740 citizens voted on how to spend €20m in the first round, and another €400m of spending will be decided on by 2020. New York City is into its sixth year of citizen consultations (now involving twenty-eight city councils up from four in 2011) – the biggest participatory budget in the US.

The moment I wrote that last sentence an incoming tweet told me of Madrid's new €60 million participatory budget. In the UK, Scotland leads the way, with successful budgets across the nation, notably in Glasgow (the funds on offer are not large but central government is keen to explore the idea going forward, providing support for twenty local authorities to embrace PB). And Portugal has announced a nationwide participatory budget, allowing citizens to submit ideas for projects that are then voted on by their peers (using, it is hoped, ATMs).* Today it's estimated over 1,700 local governments in more than forty countries are using the process to one degree or another.

So, what does all this data tell us? Do cities that embrace participatory budgeting have better outcomes than those that don't? The answer appears to be a resounding yes, *but only if they keep at it.* A 2008 World Bank report found that participatory budgeting contributed to a greater reduction in poverty in the municipalities that used it compared to those that didn't, noting this occurred even when there was a reduction in GDP per capita – in other words, *even when economic times were tough, poverty still declined.* Importantly, however, this only happened in places that kept faith with the process, revisiting it year-on-year. Municipalities who dabbled saw no real benefit.

Another heartening study (particularly if you're a parent) covered the years between 1991 and 2004 and concluded that cities using participatory budgeting showed a 'pronounced reduction in infant mortality', with baby deaths down by up to 12% over and above non-participatory municipalities – all thanks to citizen's prioritisation of healthcare and sanitation investments. There's more. Taking in data from 20,000 participatory-funded projects from around the globe, representing over $2 billion in spending, a 2014 report by the International Institute for Environment and Development found participatory budgeting projects were often 'cheaper

---

* Portugal has pedigree in PB innovation. In 2015 the budget in the seaside town of Ovar achieved 25% participation using gamification techniques (in particular an interactive leader board of proposed projects).

and better maintained because of community control and oversight' and that the process helps 'establish or rebuild trust and dialogue between people and local civil servants and politicians'. Maybe that's because participatory budgeting, by reinforcing transparency and accountability, is the natural enemy of democracy's biggest corroder – corruption.

A constant refrain from elected officials who object to the idea of participatory budgeting is that it means giving up control to a bunch of amateurs (i.e. voters). During my chat with Eduard he'd told me, 'The truth is many politicians didn't like the participatory budgeting process – it takes power from them.' But politicians take note. Embracing participatory budgeting is *good for your career*. Honestly.

When previously popular Chicago alderman Joe Moore found himself losing touch with the electorate, only narrowly winning his formerly safe seat in a 2007 run-off, he decided to bring participatory budgeting to bear on the $1.3 million he personally had the power to allocate, the first landing of Brazilian-style PB in the USA. After several months of citizen deliberation the money was put to use fixing pavements, installing bike lanes, street lighting and building community gardens throughout the city's 49th Ward. Even though less than 3% of the ward's population took part in the process, when election time came round again voter turnout increased by 23% and Moore's share by 20%. He reportedly said participatory budgeting was 'the single most popular thing I have done in twenty-two years as an elected representative.'

Yale University's Paolo Spada has trawled through years of Brazilian election statistics and discovered that parties implementing participatory budgeting raised their chances of winning the next election by more than 10%. In an echo of the views of those protesters who came to support Olívio Dutra's tax changes in Porto Alegre, the Inter-American Development Bank (Latin America's 'development' bank) found participatory budgeting increased local tax revenues 'denoting an effective increase in citizens' willingness to pay taxes'.

In summary, if you're a politician, participatory budgeting can increase your popularity, get your voters to pay more tax and save babies' lives. Politically speaking, it's not far off magic.

Duval has generously agreed to take me on tour of one of Belo Horizonte's funkier neighbourhoods. As we move from bar to bar, supping beers and munching on selections of *pastel* (a kind of Brazilian samosa, variously filled with meat, vegetables or cheese), I'm pondering how such small numbers of people taking part in participatory budgeting processes can be having such a profound effect on their cities. We stop in a public square where a crowd is watching a local pop band, all part of a citywide music festival. I estimate the audience to be 300 strong and muse that it would probably have to be twice that size to have a good chance of including someone involved in the city's participatory budget. So, how participatory is it, really? Then I realise I've got my perspective all wrong. How much bigger would the crowd have to be to have a good probability of including an elected official? In Chicago's 49th Ward one man used to decide how to spend $1.3 million. Now those same decisions are made by over 1,500 people – three times the number of Representatives and Senators in the US Congress.

Similarly, you could argue that only a portion of the budget is up for grabs in these processes, but it's considerably more than was available before and is therefore a significant step in the right direction. Participatory budgeting might only involve a small percentage of the electorate deciding on sometimes bijou sums of cash, but it is several orders of magnitude more democratic than any other form of government you're likely to come into contact with. Every dollar allocated by PB is a dollar spent more democratically than one distributed using a centrally decided budget.

Lastly, while it may only engage a small subset of the citizenry each year, over time there's evidence to suggest the process reaches many more. The World Bank study into Porte Alegre's experience found that by 2006 around 20% of the city's residents had participated in the budget at some point in their lives. 'It's a bigger interface,' says Duval. 'The people and the politicians have to grind together, and even though that's time consuming and the process is more expensive than a bunch of government suits deciding the budget, the results are far better. You've seen that.'

That night, with the never-ending hub-hub of the city's streets leaking into my hotel room, Duval's use of the word 'interface' rattles around my head. He's right – participatory budgeting is a way for citizens to plug into the business of government if they want to. They don't have to, and on balance most of them don't, but the fact the interface is there is reassuring. I try to think how participatory budgeting would work in my own neighbourhood of New Cross, south-east London. To be honest, I probably wouldn't get involved. I've a business to run, a book to finish and the demands of family to attend to. I'm busy. But here's the thing. I'd have much more faith in how the council budget was spent (and in our councillors' performance) if I knew a diverse set of my neighbours were engaged in a process like the one I've seen here in Belo Horizonte. Take note, local political door-steppers: I'd also be much keener to vote for *any* political party ringing my bell who told me their campaign involved establishing an ongoing participatory budgeting process. Ubiratán de Souza, a former political exile and the man Porto Alegre's mayor Olívio Dutra put in charge of overseeing the city's participatory budget, writes that in PB the citizen:

> 'ceases to be an enabler of traditional politics and becomes a permanent protagonist of public administration. The participatory budget combines direct democracy with representative democracy, an achievement that should be preserved and valued.'

Hollie Russon Gilman makes the point that, given the effort it takes, participatory budgeting 'can be said to represent an unlikely exemplar of innovation' requiring 'significant resources from elected officials, community-based organisations, and citizens'. In a world where digital technology is held up as the answer to nearly everything, where I find myself saying 'surely there's an app for that?' at least once a week, participatory budgeting's old-school, town-hall, get-on-a-bus, committee-meeting approach is, I realise, exactly what I like about it. Churchill may have been right about the best argument against democracy being 'a five-minute conversation with the average voter' but participatory budgeting is a door you can walk through, if you choose, to become an *un-average* voter. By the same token, politicians who walk through that door have to be prepared to get down and dirty with their constituents, with real money on the table – and then make sure what's agreed is actually delivered. Unsurprisingly it seems to make them better at their jobs, and more popular as a result.

Participatory budgeting is perhaps the ultimate, if the most prosaic, example of what I set out to find when I started this journey: a true institutional innovation. In a world where the political response to democracy's problems seems to be to polarise further to the right or the left, participatory budgeting acts as a much needed counterpoint, bringing people together. In their book *Bootstrapping Democracy*, Gianpaolo Baiocchi, Patrick Heller and Marcelo Silva describe participatory budgeting as a 'paradox of democratisation within a dysfunctional polity' – because political scientists have a habit of saying profound, surprising and hopeful things in ways that makes you feel slightly deflated and excluded. Olívio Dutra, reflecting on his time as mayor, put it better: 'Democracy's problems', he said, 'are solved with more democracy'.

I'm spending my final day in Belo Horizonte with Maria Nahas enjoying the city, moving from gallery to coffee shop and talking about her work. I am a convert. I can honestly say that I've become a believer in, of all things, a budgeting process.

Maria tells me that the unofficial neighbourhood of Capitão Eduardo where I'd seen the '*Gato*' energy system will be able to take part in the city's participatory budget next year. Even though it doesn't exist legally, a loophole has been found (partnering with a recognised neighbourhood), perhaps allowing the people I met there to finally get paved streets and a reliable power source. Maria won't be involved, however. Instead she'll be finishing a book on 'municipal indicators' and then applying her skills to help Brazil meet the UN's Sustainable Development Goals. She'll be working for the Special Rapporteur to the United Nations Human Rights Committee.

'I'm a lucky person because I have had the opportunity to live this life,' she tells me over lunch. 'The resistance, fighting for democracy and human rights, then and now ...'

We talk about recent demonstrations across Brazil, citizens taking their anger over corruption and the high cost of public services on to the streets.

A broad smile breaks across her face. 'I wish I was with them! I am them! I support them! Yes! They are *right* to demonstrate! If I were younger, you know …' In her heart, she's still a revolutionary. 'We still have a long way to go, don't you think?'

I'm coming to the end of my journey and, I'll be honest, I'm tired. But Maria's company is refreshing and revitalising and I realise that what I find most appealing about her is that she's always looking forward, and still has a fire in her belly. Like all the innovators in these pages her passion is the future, and finding a way to make it better – for everyone. Her city could also be her motto because, if you speak Portuguese, you'll know that *Belo Horizonte* translates as 'Beautiful Horizon'.

12 WORST SCHOOL IN THE COUNTRY

'You may leave school, but it never leaves you.'
– ANDY PARTRIDGE, FROM THE XTC SONG 'PLAYGROUND'

I'm trundling through the northern English countryside on the way to the last stop on my journey to hear a story that few would believe. Accompanying me is David Price OBE. OBE is short for 'Officer of the Order of the British Empire', one of a set of an awards recognising services to the arts and sciences, bestowed on worthy UK citizens twice a year by the British Royal Family. David seems a little embarrassed by his royal gong, awarded for 'services to education', telling me that what OBE really stands for is 'Other Buggers' Efforts'. It turns out he keeps his OBE in the downstairs toilet. 'I didn't know where else to put it,' he says, his north-east accent still firm, despite having left Jarrow nearly forty-five years ago.

I'm in David's company because I've been looking for a guide to help me investigate our education system, its various problems and consider how we might build a better one. It is perhaps the most emotional, controversial and tricky topic in popular political debate. Nowhere else have I come across quite the levels of passion, bitterness, anger, frustration and antagonism that typifies exchanges between those who hold differing views about what we should do with our schools. Maybe this is because our own educational experiences leave such a strong mark on each of us – a major part of the crucible in which our emotional characters are forged. Unlike the other systems I've covered in this book, education is something nearly everyone has a deep and personal experience of. As creativity expert Sir Ken Robinson says, it's 'one of those things that goes deep with people ... like religion, and money ...' – an observation made in the most downloaded TED talk of all time (40 million views and counting) called 'Do Schools Kill Creativity?' The reaction to Ken's ideas tells you everything you need to know about the tone of the education debate. On the one hand, he's hailed a 'visionary cultural leader' who offers 'a brilliant and compelling vision for what education must become'. On the other, he's accused of being 'profoundly, spectacularly wrong' and a man who indulges in 'blithely uttered rubbish' that's 'an extremely corrosive and destructive influence'. See what I mean?

Despite the vitriol and emotion there are two things that everyone does seem to agree on: a) our schools need reforming, and b) how and what we learn is inextricably tied to our future life chances and prosperity, not just as individuals, but also as organisations and nations. One of David's favourite observations is that of business theorist Arie de Geus, who said, 'the ability to learn faster than your competitors may be the only sustainable competitive advantage.' But wading into the education debate is not for the faint-hearted.

David, therefore, is something of a find. Other educational experts and commentators I've spoken to clearly have axes to grind, childhood demons to exorcise or ideological positions they're desperate to defend. David has his views, of course, but whereas a good number of the smart people I've met on this journey have wielded their intellects like rapiers, David's challenges and observations are more akin to an invite for dinner – it's all about mutual understanding and collaboration. It's an attitude that probably stems from over forty years in the education business, working at every level. David isn't an armchair theorist. He's spent decades at the coalface. Today, as an independent writer and consultant, he spends his time seeking out great learning environments across the globe in both the private and public sectors and helps his clients (governments, corporations and schools) embrace their lessons. His book *OPEN: How We'll Work, Live and Learn in the Future* is a fascinating and entertaining read about how we can all, as individuals and organisations, learn better.*

I've asked David, with his international perspective, where in the world I should go to see an education system fit for the new century, ready to graduate citizens able to address the challenges I've witnessed on my travels – and one that all sides of the debate might agree on. His answer, to my surprise, is one of the toughest housing estates in Britain and a school called Hartsholme Academy.

In 2008 no one in their right mind would have bet that Hartsholme, a school for 4 to 11-year-olds, would become a beacon for educational excellence. Following an inspection by

---

* The book begins with the story of David suffering what he thinks is a heart attack while stuck in traffic jam, but actually turned out to be a bad episode of atrial fibrillation, a cardiac misfiring that his doctor told him, in certain cases, can result in 'sudden unexpected death'. David writes about the online patient forums that helped him find the best care and so he's delighted to hear about my visit to PatientsLikeMe in Boston: 'It's the natural progression – how learning morphs from expert hierarchies to networks of expertise, which means innovation can come from surprising places.'

the UK Office for Standards in Education (Ofsted) it had been ranked pretty much the worst school in the country and placed in 'special measures' (a UK government classification which roughly translates as 'beyond bad, with staff who seem incapable of making it better'). It suffered from 'a significant legacy of underachievement', and 'a lack of leadership at all levels'. Fourteen headteachers had passed through its doors in just seven years, none of whom had been able to do much. Now almost certainly destined for closure, local residents put an advert in the teaching press with the plaintive cry, 'Please Save Our School'. This turned out to be the catalyst for a surprising and heartening transformation. In fact, two years later Hartsholme had not only achieved Ofsted's highest rating ('Outstanding'), but was being hailed as one of the greatest exemplars of educational practice anywhere in the world. (That rating is of little consequence to the man at the centre of the change, however. 'I know plenty of pretty awful "Outstanding" schools,' he will tell me.)

'So, the thing about Hartsholme', says David as we order teas from the drinks trolley, 'is that it achieved this without changing a *single* member of staff. If you look at most of the other schools that get talked about as 'world class', you'll find they were either new schools who got to recruit the staff they wanted at the outset, or they were transformed, in part, by bringing in new talent.'

By example David proceeds to tell me the story of his friend Larry Rosenstock receiving a call from President Obama. Larry's the headteacher at San Diego's High Tech High. A group of thirteen superstar schools, whose students are chosen by a blind zip-code lottery, High Tech High achieves 96% college entrance for its 5,000 pupils and is regularly hailed as one of the most progressive educational establishments in the world, turning out creative, curious, self-starting students. The President wanted to know how he might scale that kind of success nationally. 'You can't', was apparently Larry's response. 'It took me decades to get this team together.' (The schools also received significant funding from Bill and Melinda Gates, investment that is hardly repeatable nationwide.) 'I want to be remembered for the quality of my schools, not the quantity,' Rosenstock says.

'But if Hartsholme can come from the worst possible start to where they are now, and do it without firing anyone, it means their success is likely to be *replicable*,' says David. 'That's the reason we're on a train to Lincoln and not a flight to San Diego.'

In other words, Hartsholme needn't be an isolated success story. It could offer a template for creating a much-improved school system and, as I shall find out, one that, miraculous though it seems, might satisfy everyone.

A useful metaphor for understanding the debate in education can be found by looking at the history of Western music – and what David Price doesn't know about music may not be worth knowing. Something of a musical prodigy, he left school at 17 to pursue a career in the entertainment business (via a brief and disastrous stint as a clerk for the Ministry of Pensions and National Insurance). By 21, he was writing songs for other artists of the day, scoring some minor hits, and his band Cold Comfort were touring their debut album recorded for Jet Records, home of the Electric Light Orchestra and, later, Ozzy Osbourne. It was Osbourne's future wife Sharon who ended the party, after looking at the sales figures. 'I don't blame her,' says David. 'ELO were raking in the money, and we were nothing by comparison.' This was the prompt for David to enter the world of education, where, by his mid-thirties he was a much-loved director of performing arts at Manchester

College of Arts and Technology, creating environments to fuel the creativity of a generation within the city. In 1994 he was asked to help establish Paul McCartney's now famous Liverpool Institute for Performing Arts (based in the Beatle's old school), largely designing the curriculum and becoming the institute's Director of Learning for the first eight years. Following this he was hired by the Paul Hamlyn Foundation (one of the UK's largest grant-giving charities) as director of the Musical Futures project, created with the intention of getting kids more interested in classical music. His first recommendation, much to the consternation of his new employers, was to remove the focus on classical music.

'I had to explain to them that for most kids classical music is a bit like gardening – you grow into it,' he says as we approach Lincoln station. 'Instead of saying, "This is classical music and this is why you should learn it", you start with where the kids *are* and ask them how *they* want to learn music. That can be challenging for teachers. They say, "but the kids will choose hip-hop or some music that I don't know anything about!", to which my response is, "Well, you have a pair of ears, don't you?" A music teacher shouldn't be scared of learning new music. You've got to bridge the gap by coming at it in both directions.'

Studies by the Institute of Education have found that not only does the Musical Futures approach increase students' enjoyment of, and achievement in, music lessons, but it also has a positive impact on the whole school, improving kids' motivation, well-being, self-esteem, confidence, organisational skills, concentration and their ability to work in teams. Teachers using the approach become more confident and effective with their classes too – all reasons why the approach has now spread to thousands of schools internationally. In short, it's a win for pupils, a win for teachers and a win for schools. (And if you're the working-class boy from Jarrow who oversaw the whole thing, it'll get you an OBE.)

Making generalisations about music is, of course, a fool's game but part of the divide that's arisen between classical music and other genres has its roots in the ninth century, when monks started writing down their chants using simple musical notation. Initially the blobs on the page only told you if the next note was higher or lower than the last one, but over the centuries a number of innovations slowly coalesced into the musical notation we use today – a complex set of dots, lines and short instructions in Italian which tell you the pitch and length of each note (as well as its relative volume) along with the tempo and overall dynamic structure of the music – meaning if you can 'read music' you don't have to listen to a composition before you can play it.

For many people, notated classical music came to represent a 'high art', variously referred to as 'serious', 'erudite' or 'legitimate' music (whose most celebrated exponents train in the rarefied confines of a 'conservatoire'), while other traditions, where musicians played without written music or enjoyed the off-notation thrill of improvisation (often grouped under inexact headings like 'folk', 'jazz' or 'popular' music), became considered 'lower' art forms.

'So, one outcome of this divide is that there has become a "right" way to play classical music,' says David. 'The notation is exact concerning what notes go where and the window of interpretation is pretty narrow.' (Indeed, the term *conservatoire* is derived from the word 'conserve'.) Other genres of music allow a looser take, meaning performances can vary and that's all part of the enjoyment. It should come as little surprise, therefore, that when children want to learn how to play music (and lots of them do), some can find the classical approach, with its complex codified language and associated strictures, less than engaging, because it takes a lot of the play out of playing – something David and his colleagues aimed to fix with their Musical Futures approach.

I have a feeling that it's David's deep understanding of our two parallel musical traditions, their merits and challenges (and how to use them as gateways to each other) that helps make him a constructive voice in the larger education debate, which can be broadly characterised as having two similarly competing modes of thought. On one hand are 'traditionalists', who, at their most traditional, favour direct and guided instruction in discrete topics from an expert and respected authority (the teacher at the front of the class). Long-fought-for knowledge is transferred to you for your benefit. Like the staunchest guardians of classical music the traditionalists are clear there are obvious boundaries between what's worth our attention (often 'high culture') and what's not. What's more, as in classical music, the 'right' way to do things has been long established and it therefore makes a great deal of sense to learn it if you want to get on in life.

In the opposite corner are the 'progressives', who, at their most progressive, believe every learner's journey should be personal, guided by their passions and interests (rather than a mandated curriculum of chosen and separated subjects), where teachers are not authority figures but guides who model good learning behaviours. Progressives talk less about straight knowledge transfer and more about skills development, particularly collaborative problem-solving, believing learning is as much about experience as instruction – it's a group jam, not a watched performance.

Neither of these positions (and the myriad gradations in between) is new (we've been arguing about the best way to educate for years) or, crucially, mutually exclusive. What makes the debate so heated is the feeling that, as the world becomes increasingly complex, *we really need to get this right*. For the uber-progressives the problems we face, whether it's climate change or inequality, are partly the result of a 'traditional' educational approach that they say hampers us from thinking creatively across subject boundaries – and they therefore argue a fundamental rethink of the whole system is needed. For the arch-traditionalists it's the progressives' abandonment of rigour and their lack of deference to expert authorities and discipline that's at the root of our current problems, and what we need is to 'get back to getting the basics right'.

What's a government, charged with overseeing the nation's education, to do? It's in considering this question we see why traditionalists have had the upper hand in shaping educational policy in most countries. Governments need to know if they're improving things education-wise. To do that they need statistics. The trouble is, measuring how well educated someone is isn't easy. How does one assess, for instance, an individual's ability to think creatively or ask the right question? How can we evaluate someone's capacity for empathy or their ability to collaborate? There is no simple answer. But we *can* measure, without much difficulty, whether someone knows a fact or not by sitting them in an exam, separated from everyone else, and asking them to recall it. By the same token, skills that aren't dependent on social interaction, such as mathematical problem-solving, can be assessed in a similar manner.

And so nations create curriculums skewed toward facts and skills that can be *easily examined*. Students, parents, teachers and nations, keen to score well, become fixated on exam results – a culture which inevitably favours a traditionalist approach to education, sidelining harder to examine 'soft' skills, and keeping the progressives on the back foot. It's also worth bearing in mind that the people in power have been through an education system like the rest of us, but it worked out alright for them – after all, they've been pretty successful – which may also tip the balance in favour of the traditionalists.

At its worst this approach sees cramming more facts into students' heads the ultimate measure of success and, as the fact-burden increases, educators may find themselves increasingly 'teaching to the test', believing it's the only way to help their pupils succeed in the system. Taken to its extreme (like most things taken to the extreme), this can have devastating consequences. Uber-exam-obsessed nations like China and South Korea are suffering an epidemic of student suicide because exam results have become such a totem of personal worth. Any measure of failure is literally intolerable. It's so bad that some schools now install barriers around exam rooms on higher floors to stop stressed pupils jumping to their deaths before important tests, while some of the universities that students compete to get into are asking new entrants to sign 'suicide disclaimers' absolving the institution of any responsibility should they kill themselves. Luckily things aren't that bad in most places, but it's still a common argument that our schools have become little more than factories, trying to produce standard products (students) while the workers (teachers) become increasingly demoralised by the regimented nature of their profession. This broadly traditional approach can suck out the joy of teaching for teachers and learning for students. Something, after all, has to explain why anything between 30% to 60% of students are 'chronically disengaged' by the time they reach high school (a figure that doesn't include those that have already dropped out – about 7% of students in the US and UK).

The consequences of an aversion to school are particularly serious for children from disadvantaged and poor backgrounds, like those David and I are on our way to visit. They don't get the 'second chances' that kids from more affluent households might enjoy, and their disengagement, according to a report from the US National Research Council, 'increases dramatically their risk of unemployment, poverty, poor health, and involvement in the criminal justice system.'

That's not to say all kids hate school. The results of a 2012 Canadian study of 63,000 schoolchildren are typical, finding that, while only 39% of them found lessons engaging, 69% were engaged with the social aspects of school and, importantly, the idea of school – understanding that it's there to help them improve their life chances. In short, kids understand that school is a good idea in principle but find lessons dull.

So, what's to be done? Every parent wants the best education for their children, every learner would do better if they were genuinely excited by their lessons, employers wish for a well-educated and increasingly creative workforce and every nation hopes for a populace able to learn, innovate and collectively prevail as the world changes around us. Yet we have education systems that, whilst not hated, are missing a huge opportunity.

Is it possible to create schools that our kids beg to go to, which deliver to both the traditionalist and progressive agendas? And, perhaps even more implausibly, could an ex-nightclub owner hold the answer to the conundrum?

Carl Jarvis is happy to see you. It doesn't matter who you are, he's genuinely glad you're here. From the moment you're in his company until the moment you leave you get the very real impression he finds you interesting and insightful. This attitude, I shall find out, is one of the reasons he's become considered something of a miracle worker. Tall, clean-cut, with short-cropped but slightly unruly white-blond hair, and cheeks that look well exercised from smiling, his manner is the very definition of welcoming. Chatting to him in his office at Hartsholme Academy, he is like your favourite pub rendered in human form, oozing an authenticity that immediately makes you feel comfortable and liked. It's a persona I find all the more surprising when I hear his personal story.

Born into a working-class family in Nottinghamshire (his father was a miner, his mother a cleaner), Carl hated school and, he remembers, school didn't much like him. 'One of my biggest memories is a teacher talking to another adult about my inability to read and write, aged seven, and this other person saying, "he doesn't need to, he's going down into the pit, so leave him." Even at seven I realised where I was in the pecking order; way down. They thought I wasn't worth an education.'

Tragically, the young Carl internalised the message and by the time he entered his teens was 'a complete nightmare, one of the worst children in school. My crowning moment was smashing the fish tank in reception. Thousands of gallons of water ran into the gym and destroyed the wooden floor. It was a trail of devastation. Fish everywhere.'

At that moment few people would have bet on Carl becoming one of the most respected educators in the country, although one teacher did see potential in the alienated tearaway. 'Mr Cox, my English teacher, took me under his wing. I don't know why. Maybe he was a fish lover, but he became my guiding light. He found the thing in me that would motivate me to learn again.'

'Which was?'

'I wanted to join the Air Force. He made me see I wouldn't make it if I didn't study. He helped me believe in myself again. He was always raising my own expectations of who I could be. I worked jobs outside of school, washing pots until midnight every night to pay for flying lessons – and got my pilot's licence aged 17.'

It was a massive turnaround, thanks to one teacher who'd found the hook to engage the teenager. Without that input Carl wouldn't have found himself, aged 18, at the end of a gruelling week-long selection process for the Royal Air Force, down to the last three candidates from a carefully selected intake of fifty. 'So I do the final test, a leadership test, and I'm told I've passed, along with two others, who are both from private schools. But the officers in charge have a problem with me. They say, "At your interview on Wednesday you told us your father reads the *Daily Mirror* and anybody whose father reads the *Mirror* cannot be an officer in the Royal Air Force."'

Carl was simultaneously devastated and furious.

'I'd given every moment of every day over to this dream – and then to be told "because of your parents, you can't have it …"' His eyes narrow, his mouth tightens. 'I'll never forgive them,' he says. Seeing his face betray the anger he still clearly feels is disconcerting, because the expression seems so out of place.

Another system and another social pecking order had told Carl he wasn't worth investing in because of his parentage. 'Dazed and heartbroken', he got a job as a manager for a bar that, not long after, announced it was closing. 'I bought the lease and the debts for a pound,' he recalls. 'I had no money but I thought I could take the risk – at least this was something I was in control of.'

By the time he was 24, Carl and his new business partner were running five nightclubs and three pubs and 'loving every minute of it'. Then, on Christmas Eve 1994, he had an epiphany. 'I was stood on the door at a club, it's snowing, and I thought to myself 'Do you know what? I'm going to sell my bit of the business and I'm going to go into education to stop people doing to other kids what people did to me – and that's it'.

However, as soon as he entered teaching his ambition hit the buffers of reality. 'What I found was this Victorian system that didn't want to change. It was *so* old-fashioned and I thought, "God! Why are we doing it this way? Surely we can do this better?" And that's when I started all my research, and trying to find out everything I could to make a better system.'

Carl, previously the most disenfranchised pupil in school, was now a studying machine supplementing his experience in the classroom with explorations into psychology, neuroscience, theories of innovation, communication skills, corporate research into teamwork – anything he could get his hands on. He rose quickly through the ranks, becoming a headteacher within a decade, and gaining a reputation for innovation coupled with good management – a safe pair of hands who was also something of a maverick. 'I'd got myself the job of leading a lovely primary school in a very nice area and then I saw this advert – "Please save our school" – and I thought "That's interesting, I have to take a look."' Recalling his first trip to Hartsholme he describes it as '*completely* broken'.

'It was absolutely the worst school I'd ever, ever been to. The children were feral, running over desks, teachers shouting across rooms trying to get their attention. As I walked through the front gates there was a child on top of the gate, with staff literally begging him to come down. It was mayhem. If you asked the kids what their aspirations were, some would genuinely answer they wanted to go to prison because that way the State would look after their mum. I was looking at the staff and it seemed to me they'd given up. It was a genuinely scary place and I remember thinking there was no way it could be saved. I was like: "No thanks".'

But on the way out, he changed his mind. He'd been written off twice because of where he was born and to whom. Was he about to let history repeat itself? Hadn't he got into teaching 'to stop people doing to other kids what people did to me? I got to my car, turned around, looked at the school and thought to myself, "You know what? If I don't do it, nobody's going to do it. Those kids deserve a decent education and I'm going to give it to them."'

He took the job.

On my walk around Hartsholme Academy today it's hard to imagine the place Carl describes from his first visit. The place is buzzing, not with the threat of uncontrollable children kicking-off, but with an air of friendly industry. David, Carl and I are in a classroom with no chairs, decorated in a space theme. Planets are hanging from the ceiling, there's a mock-up of a Space Shuttle inside which some eager astronauts are working, and the lights are off ('It's dark in space!' one of the class explains). It's all part of creating an 'immersive environment' co-designed by the children to reflect the big themes they choose each term, themes that act as a springboard for the rest of their learning.

'You have to ask yourself why we dull our kids' senses by sitting them in rows, telling them to be quiet and not to talk to anybody,' says Carl. 'Take that away and give children an *experience* rather than a curriculum and they learn ten times more, and so much faster.'

Appropriate to their age, this class will explore the science, history and human stories of spaceflight, learning all the curriculum's demands for literacy and numeracy in the process, but also getting a healthy dose of 'soft' skills too. Scattered around the class are small huddles of children, working together. 'We never tell them where to work. We never tell them who to work with. They've got the tasks we give them and they organise themselves.'

Instantly, and I can't help it, my inner traditionalist pipes up. 'But surely they gravitate to their best mates? Don't they just ... play? If they're free to go wherever with whoever, who makes sure they're learning?'

Carl chuckles. 'Yes, we were worried about that too. It's actually something we learnt from Google, who give their employees quite a bit of freedom as to who they work with and on what. And I'll be honest: when we first considered trying it here we thought the same as you – "It won't work in a school setting" – but it actually works brilliantly.'

The reason it does, says Carl, is because Hartsholme works tirelessly to nurture learners who are self-motivated, 'creating an environment where it's a matter of personal pride for them to do well. So what we find is they create groups that they're going to be the most successful in. We're clear what success looks like and they seek out the other students who will help them achieve that. What you have to realise is that children – humans – learn more effectively from their peers than they do from teachers or a book. We're social creatures and we learn socially. In what other circumstance in life, outside of formal education, do you learn by having someone talk at you? In the real world we learn mostly in teams, right?'

To make the point Carl stops a girl walking past us, head furrowed in concentration. I guess she's about 8 years old. 'Darcy. Would you mind telling us about how you help each other learn in class?'

Darcy considers this for a moment. 'Our friends help us up-level our work. So if I'm stuck on punctuation I might go to someone who's good at that.' She points to a girl across the classroom. 'Like Lily! She might help me with hyphens, or brackets, or something. Or for maths you might talk to Gwyer,' she points out a boy busy in a group of three. 'He's *really* good at maths. And when *you're* helping someone else, because *you're* good at something, you have to give *specific* help – you can't just say "that's good" or "that's bad", you have to say what bit is good, or bad, and why.'

'They also assess each other,' David chips in, 'so there's a lot of peer-motivation going on too.'

'They mark each other as well?!' I'm already writing a tabloid headline in my head: *Where's Teacher? Ultimate madness as school lets kids do their own marking.*

Darcy explains. 'First you have to self-critique,' she says – and I'm impressed with her command of language. 'That's a bit like marking yourself, where you say how you think you've improved to someone else. And then your friend marks you.'

Carl can tell I'm slightly sceptical and reassures me that 'Ultimately this is all overseen by the teacher, so there are standards they have to reach, and knowledge and skills they need to acquire. But it's just a fact that working socially is much more efficient and enjoyable for the pupils, so we give them the tools to assess and encourage each other, which is a lesson in itself about teamwork and using your judgement.'

As we wander into another classroom (this one bedecked with scenes from Beatrix Potter books), I ask Carl, 'How do they deal with the constraints of exams when they've had such a free environment?'

'Generally they're very eager to do them.'

Excuse me? Now I'm really getting incredulous. Kids who *want* to do exams? It's a traditionalist's dream, but I'm not sure even the staunchest of them would believe children actually like doing tests. 'It's not a chore for our kids,' explains Carl. 'They're keen to show off what they've learnt. It's a self-esteem thing again.'

It probably why students here consistently ace the UK government's Standard Assessment Tests (SATs). Compared to other schools of its type, Hartsholme is *top* of the UK's achievement tables, with many students officially 'beyond expectations'.[*] Remember, this was pretty much the worst school in the country before Carl arrived.

'I've got children who are five years ahead in their development,' says Carl, although you can tell he cares little for official markers of where his pupils should be. 'Those formal milestones are a nonsense really, and we don't think about them. They're not "gifted" – they've just been given opportunities. What we do is create an environment where everyone can achieve their potential.'

And he means everyone. There's no hiding it: Hartsholme's immediate neighbourhood is a tough one. There are high levels of unemployment, poverty and crime. Nine months after Carl arrived, a local arsonist actually burnt down most of the school. 'Luckily' this was in the summer holiday and the newly motivated staff gave up their time to knock down what couldn't be saved and refit what could, ready to reopen for the autumn term.

'When I arrived, drugs were high on the agenda in the neighbourhood,' recalls Carl. 'There was violence with baseball bats in the playground. It was that bad.'

Today, the neighbourhood has improved slightly, mostly thanks to the impact the school under Carl's leadership has had on the area, but it's still fair to say that this part of Lincoln is a long way from being a suburban paradise.

'This year 70% of our starting kids turned up on their first day with almost no real ability to speak at 4 years old, and most of

---

[*]   In the arcane world of the UK's official grading of schools' performance, 100% of Hartsholme's students leave having achieved 'Level 4' in reading, writing and maths, something only 75% of students are expected to do nationally. In fact, in maths, writing, reading and grammar, 46%, 54%, 60% and 69%, respectively, are hitting 'Level 5'.

them were still in nappies, and we've got a high proportion of special needs kids,' says Carl, using an education sector catch-all term that describes children who have medical, mental or psychological challenges. 'The expectation in most places is that those pupils will do worse because of their starting point. A lot of schools put those kids, with all their different needs, in a single class, which is insane. Also, once you put a child in a box that has 'unlikely to achieve' on the lid, they *believe* it. So we put them in the same classes as everyone else, and we – by which I mean us as staff, but also the other students – help them overcome their challenges. They generally do as well as their peers. We get sent kids from other schools who have been diagnosed with ADHD, but that disappears here.'

The traditionalist in me starts to complain again. I've grown up with the belief that it's important to stream classes by ability and need so that, depending on each child's individual talents and challenges, they aren't either being held back or left behind. The evidence, however, is on Carl's side. According to Ofsted, 100% of Hartsholme's 'disadvantaged' pupils are reaching the standard expected for their 'un-disadvantaged' peers, with 20% of them exceeding it. In fact, it turns out that streaming kids by ability has an almost negligible impact on students' performance whatever school they're in. It seems that a lot of the ideas many of us have about how to run an effective education system are misguided, no matter how intuitive they appear.

In 2008 John Hattie set the rational cat amongst the prejudiced pigeons. Professor Hattie upset traditionalists and progressives alike by, ironically, telling them they were both right.

Hattie is a meta-researcher like Dr John Ioannidis, the man who exposed so many flaws in the practice of medical research I'd learnt about during my investigation into PatientsLikeMe. For the last twenty-five years he's been collecting every study he can find on student achievement, analysing it and trying to find patterns. He's amassed data on over 195 influences on our children's education, from the effect of uniforms to the impact of various teaching philosophies – culled from over 70,000 studies spanning 50 years and involving over 250 million students. It's by far the largest and most ambitious analysis of education research ever undertaken. Hattie admits, almost by definition, it must be approached with some caveats in mind, not least because he's had to make judgements on the quality of each study and draw out patterns across research done at different times, using different methods, in different schools and concerning different numbers and mixes of students (and inevitably there has been some heated debate about the way he's gone about this). As he says 'a common criticism is that [meta-analysis] combines "apples with oranges" [but] in the study of fruit nothing else is sensible.' His analysis, he admits, is not gospel – and he's also quick to stress that it doesn't deliver a 'top ten' of approaches that should be applied in every school, because every school is different. What it does provide, however, is one hell of a starting point to discuss the 'big picture' in education and perhaps begin to draw some useful conclusions.

Hattie graded each of the 195 influences by 'effect size' – a somewhat blunt and subjective statistical technique usually ranked between −1 and 1. If an 'effect size' is below 0, whatever you've done has had a negative impact, while anything above 0 signals a positive influence.

So, what has Hattie concluded? In short, nearly everything works. In fact, '95% to 98% of things we do in the name of enhancing achievement does enhance achievement', he says. By Hattie's analysis the average 'effect size' across the 195 influences is 0.4, which he uses as his notional benchmark. Anything with an effect size below this is hardly worth shouting about. Interventions that deliver an effect size of 0.3 are, of course, having an effect, but in the pantheon of things you could do to improve student achievement they are by his definition 'below average'. The problem, he says, is that we tend to put our efforts into these interventions, while ignoring those where the impact is much larger. Our priorities, he argues, are desperately skewed.

By example, consider the topics that get the most airtime in discussions about education: increasing school funding, reducing class sizes, streaming kids by ability or coming up with new types of school administration – the 'structural things', as Hattie calls them. The average effect size of such interventions is an underwhelming 0.1 (none get above his average of 0.4) – a surprisingly small impact. They dominate education policy even though 'they don't matter much', suggests Hattie, because, like exam results, they are things we can see.

Saying this kind of thing annoys people. Traditionalists may not warm to his assertion that 'if a school is debating a uniform it's in big strife. It's created a massive distraction. There is a zero effect of uniform, which means I don't care whether you have it or you don't, just don't make it an issue.' People on both sides of the progressive/traditional divide are also outraged to hear Hattie say that 'the stupidest, craziest, inanest, puerile thing we do in education' is to obsess about reducing class sizes, something both groups generally agree on. It's not that reducing class size doesn't have a positive effect (0.2), but compared to other interventions it shouldn't be as far up the agenda as it is. (The reason for the low effect is, according to Hattie, that teachers teach in pretty much the same way whether they have a class of 40 or 15.)

Just to be clear, Hattie is not saying we should ignore getting the structural things right (he's in favour of smaller class sizes), but that holding them up as the key route to improving education is like trying to improve an old car's performance by pumping up the tyres while ignoring a much-needed engine refit.

So, let's move on to the things that fuel a good deal of the vitriolic debate between the progressives and the traditionalists – the philosophies by which we teach. Should we have teaching based around problem-solving, for instance? Do we need 'individualised' (tailored) or 'programmed' (one size fits all) instruction? Turns out it doesn't make much difference. Both have an effect size of 0.23. What about using technology – introducing computers, tablets and Internet tools into lessons? Surely that must improve things? In fact, not very much – an average effect size of 0.22, an impact that hasn't changed much for fifty years. Technology, says Hattie, is a revolution in education that has always been 'coming', but has never arrived.

If you accept these findings, you can understand Hattie's frustrations with a system that seems to ignore a lot of things with large effect sizes simply because they don't fit with a particular educational ideology. That peer-tutoring I baulked at? Hattie gives it an effect size of 0.55. I didn't like the sound of it when I first heard about it, but I was wrong.

Hattie hadn't published his findings when Carl took over Hartsholme, so it's another testament to the frustrated pilot and former nightclub entrepreneur that his own research and experiences had led him to many of the same conclusions about what works and what doesn't. This meant that on day one he had a strategy for Hartsholme, but it wasn't one his new employers, the local government authority, were all that keen on.

'They asked me to get rid of the staff,' says Carl. 'I was told, "They're all rubbish, sack the lot of them."' He resisted, although not because he'd received an enthusiastic reception from his new workforce. 'You can imagine their attitude. I could tell they were thinking, 'Here we go again, he won't last long.''

What happened next has become the stuff of legend at Hartsholme. Carl sent each teacher to their classrooms with instructions to bring back all their lesson plans, along with every bit of curriculum guidance they could lay their hands on. Expecting on their return to be asked detailed questions about their approach, they instead found their new boss standing behind a large bin, into which he asked them to throw everything they'd retrieved 'because, let's be honest, it's clearly not working'.

Then he told them that within two years Ofsted would be ranking the school 'Outstanding'. This ambition was met with outright laughter, but Carl was having none of it. He'd seen that both the students and the teachers at this school had the *same* problem, and one that could be solved to their mutual benefit.

John Hattie's research concludes that two of the most powerful influencers on student achievement are: the interplay of what a learner expects to accomplish, and their teachers' assessment of the same (huge effect sizes of 1.3 and 1.6 respectively). In fact, these two sit in the top three of his 195 influences on achievement.

It works like this: if you think you're going to get a 'C' and your teacher agrees with you, that's almost certainly what you'll get. But if your teacher tells you you can get a 'B' and believes in your ability to defy your own expectations, then you often get the higher grade. Carl was pulling the same trick on his disillusioned staff, just as Mr Cox had with him. While the teachers at Hartsholme expected at best to avoid closure, Carl was telling them they would become one of the best schools in the country – and that he personally believed in them.

He spent weeks tirelessly observing and encouraging his teachers. 'I told them they were all amazing, all the time. Even if the teaching I saw was *terrible*, I would pick on some small thing that was OK and praise it. I went over the top, but I had to, because I had to get them to believe in themselves again. I spent the first six months not in my office but in classrooms, watching things get better.'

Simultaneously he instituted the same peer-teaching and assessment approach advocated in the classroom among the staff. Carl realised his teachers, like many in the profession, had become atomised. They didn't collaborate or give feedback on each other's work. They never saw each other teach. They didn't discuss the impact they were collectively having on students and how they might work better together to improve it. In short, and with no small dose of irony, they were teachers who had *stopped learning*. They weren't acting as an organisation but a set of individuals. John Hattie says this is not uncommon. 'Too many teachers believe the essence of their profession is autonomy. We hardly ever get together and look at each other's teaching. That is a major hindrance to working collectively. I can't imagine many other professions where that happens.'

Carl started to arrange discussions amongst the staff on some of the research he'd amassed, largely staying out of the debates that followed, but watching his staff properly interact on the subject of teaching for the first time in years – and start to come up with new ideas as a result. He got his teachers to watch each other and give feedback on how they might improve, all against a constant backdrop of his high aspirations for them. He didn't know it at the time but he was working on the other influence that makes John Hattie's top three, 'collective teacher efficacy' (effect size, a whopping 1.6).

Of course, I've come across the term 'collective efficacy' recently on my travels – in Detroit, where the growth of community urban farming has been able to improve health, education and social justice outcomes through the same mechanism. What's so important about 'collective teacher efficacy' is that it has been found to be 'more important in explaining school achievement than socio-economic status'. This means that, if you can get it right, high achievement is within reach of even the most disadvantaged students, something Hartsholme amply demonstrates. You can't change the socioeconomic environment in which a school finds itself, but you can, as Carl did, change the way the staff think about that school. And it worked. Ofsted came back eight weeks into Carl's tenure, expecting to recommend the school for closure. Instead they found teachers and lessons they could now rank as 'Good' or 'Outstanding'.

'They were asking, "Is this the same school?"' says Carl with some pride. 'They said they'd never seen a transformation like it.'

John Hattie is clear that 'teachers who work together collectively and collaboratively, the ones that say "My job is to understand my impact", have the biggest effect, not the teachers who say, "It's my job to cover the curriculum, my job to get kids through the exams."' The most recent Ofsted report will tell you that Hartsholme 'provides an outstanding education for its pupils' who 'learn in an atmosphere of mutual respect and social responsibility', all thanks to 'the pursuit of excellence.' 'Pupils are happy, motivated and successful', with the staff, who show 'exemplary teamwork', making 'an outstanding impact on pupils' learning and well-being'.

And it's not only Hartsholme. News of Carl's transformation soon spread: the miracle headteacher who'd done the impossible. Soon he was working with other schools in the area (both primary and secondary) and, ever the entrepreneur, establishing Eos Education (named after the Greek goddess of the dawn) – a teacher training, schools consultancy and events company now working with at least thirty schools in the UK, and growing rapidly. Every school is approached differently (clearly teenagers are a different challenge to the under-11s) but the central principles of placing consideration of the student at the centre of all decisions, flexible classroom environments, self-directed learning, teachers working collaboratively and constant research into the latest methods remain at the core.

Hundreds more schools both in the UK and abroad, are asking for Carl and his staff to come and work with them. It's worth stressing that the teachers leading Eos are the very same people who'd 'given up' (by Carl's analysis) when he arrived in 2008. His replacement as headteacher at Hartsholme, Sara Pearson, is another of these previously doomed educators.

'These ideas, these strategies, these principles are working well across a number of schools now,' says Carl. 'It's no longer about me, and it can't be if it's going to scale. It's about creating a culture of change within the system.'

The unwavering mantra? Teamwork. Or, as John Hattie puts it, 'I'm getting a bit tired of the argument that it's about *the teacher*. It's about the *teachers*.'

And you know what I'm thinking? Is that it? I'm not taking away from what Carl, and now Eos, are doing, but really, what have I learnt here? That an organisation works best when its staff have high aspirations, and collaborate and communicate well in the pursuit of them. That students are best served by teachers who push them to aspire beyond their own expectations.

I catch up with David, who's having a chat with some 9-year-olds. 'I've got to be honest here,' I say. 'It's hardly rocket science, is it?'

'No,' says David, 'and that's what makes what's happened to so many schools so troubling. Education has become so politicised it finds it hard to reach agreement on what works. There's always some government minister, with their pet ideology, telling teachers what to do. Argument has trumped teamwork and evidence. If we were medics, we'd still be arguing about whether to wash our hands before operations. Carl only got to expose that because this school was such a disaster. He got under the wire.'

I squat down next to two 9-year-olds, Theresa and Cameron, and ask them if they like school.

'Yes!' they both reply in unison.

'And what do you like about coming to school?'

Theresa gets in just ahead of Cameron. 'The learning!'

Cameron adds, 'And it will help us get into university and get a job.' He pauses. 'I want to be a stuntman.'

Theresa, not wanting to be outdone, says, 'I want to be a doctor.'

It's too perfect, so much so that I laugh. My inner progressive and inner traditionalist are both happy.

'We wish we could come to school every day,' says Cameron without any prompt from me. 'And right now we are doing maths!'

Theresa draws my attention to the tiles at our feet, on which the sums they are working on are scribbled in chalk. 'We're trying to figure out some problems here!' she says, with a touch of impatience in her voice.

I get the hint.

13 SYSTEM ADDICT

'Defects are not free. Somebody makes them, and gets paid for making them.'
– W.E. DEMING, ENGINEER

Back in 2010 I spent a very enjoyable week in the company of double act Bruce Ward (farmer) and Tony Lovell (money man) who showed me around some farms in New South Wales, Australia. It was all part of the research I was doing at the time into a method of cattle farming analogous with the System of Root Intensification I've more recently witnessed in India's Jharkhand. The 'new' method, called Holistic Management (HM), mimics the way cattle move on natural grasslands like the Serengeti. Instead of splitting animals into small groups and leaving each in their own paddock, where, with nothing else to do, they overgraze and destroy their own food source – a farm using HM operates in tune with the animal's natural inclinations, a single large herd circumnavigating the property, staying in each paddock for only a day or two before moving on. It means most of the paddocks are empty most of the time, allowing them to recharge themselves with fresh grass, ready for the time their bovine guests return, maybe a hundred days later. In the right environment it can turn an overgrazed dustbowl into a verdant grass-fest that not only boosts business, but also pulls a healthy bounty of carbon out of the atmosphere, putting it back in the soil as more plants and grasses do more photosynthesis – a virtuous circle of increasing fertility.

Like SRI, Holistic Management is, if you'll excuse the pun, rooted in observing what nature does and tweaking it for agriculture, rather than trying to replace it with a 'better' system. The results of adopting the method were, I found, hard to argue with. On the Holistically Managed side of the fence were properties considerably more lush than their neighbours, overseen by farmers enjoying good profits thanks to higher stocking rates and lower costs. On the other side: a parched set of paddocks, declining in productivity, and farmers often surviving on 'drought assistance' from the Australian government.

After seeing a few too many thriving HM farms starkly contrasted with their 'traditional' neighbours, I asked Tony why those struggling to survive weren't leaping to embrace a system that they could see, literally by looking over a fence, was better all round. 'Ah well,' said Tony, always ready with a quip, 'these are special Australian fences. They're impervious to new ideas.'

It's a line that has stayed with me, throughout the research for this book. Reflecting on it in the light of talking education with David, I now think Tony was wrong. It would have been more accurate for him to have said that those fences were impervious not to new ideas, but old ones. Holistic Management is applying a lesson to farms that nature learnt thousands of years ago. Similarly, the understanding that teachers who learn well together create better schools is hardly a blinding or fresh insight.

There should be nothing at all radical about what Carl is doing, or what John Hattie's research has revealed. These are old ideas. In fact, when, with some trepidation, I tell my father-in-law, a semi-retired maths teacher (and ukulele obsessive), that I'm writing about his profession, he volunteers almost immediately that 'you'll find the schools that do best for their students are the ones where teachers get to work as a team, learning from and feeding off each other.'

'That's what John Hattie's research calls "collective teacher efficiency",' I say.

He's never heard of John Hattie.

Driven by a desire to measure success, it seems we've thrown out the baby with the bathwater, creating a system of education that so relentlessly chases exam results it appears to have forgotten the basics of how to run a functioning organisation. The utter obviousness of education's problems, and the current system's inability to address them, despite a large part of the answer staring it in the face, is perhaps the most stark expression of something I've seen repeatedly on my journey: systems that have failed to evolve and now exist to serve themselves first and their core purpose second, no longer the means to an end, but the end themselves.

Those structures arose in the form they did for good (or at least understandable) reasons – and achieved much. Billions have enjoyed their dividends. It is because of this legacy of achievement they expect, it seems, to rule the roost. They are no longer just systems; they have become *belief* systems about how we run society, embodied not only in thought, but in bricks, mortar, computer code, business models and the machinery of government. The people who run them are not evil, or stupid, or incompetent – rather, they are actively shielded from a bigger reality by the cultures created around them, cultures that say 'You're doing a good and important job.' Just like a teacher who can prove that pretty much anything they do has a positive effect on their students, so the employees, financiers and leaders of most organisations can point to their successes and impact while avoiding the elephant in the room – that the system they serve, in its current form, has now passed its sell-by date and might very well have become dangerous.

You may work for a drugs company whose model of development is now so expensive and inefficient it means millions die unnecessarily, but you can point to drugs made and lives saved in the past. Agribusiness executives can quote figures showing the billions of tonnes of food they've helped to grow, while neatly sidestepping the uncomfortable truth that in doing so they've helped create a system that decimates the water sources upon which humanity ultimately depends. The energy sector prides itself on powering our homes and businesses, apparently giving it license to do precious little about air pollution or climate change which threatens to destroy vast swathes of the world economy and bring untold misery to billions. It's even created the fig leaf of the 'Energy Trilemma' – 'you can't have it cheap, green and give it to everyone all at the same time' – to formalise its inability to innovate. And underpinning all this are governments and education systems that, rather than rising above the curse of systems-hardening, seem to have embraced it as a virtue.

Are we cursed to endure a zombie-like inertia as these juggernauts who forgot to keep hold of their steering wheels fail to take the sharp turns in the road we need them to, if we're to have a chance of making this world more sustainable, just, equal and humane? I hope not. It's the reason the innovators in these pages, and others like them, are so important. They hold up a mirror to our current belief-bound systems, a mirror that reflects an uncomfortable truth: 'What you believe is wrong and I have proved it ... by doing it better.'

Carl Jarvis's disdain for the existing education system may

irk some of those who work to sustain it, but the results his students achieve are hard to argue with. You can try and sell fertiliser, pesticides and irrigation equipment to a Ranchi farmer succeeding with SRI if you like, but their crop speaks louder – the harvest is an eatable (and economic) truth you cannot deny. The half-price community-owned energy the residents of Güssing have been enjoying for years may make the big utilities uncomfortable, but that doesn't make it any less real. Open Source Drug Discovery might challenge existing pharmaceutical business models, but that doesn't stop it working. (As I near publication of this book, Samir e-mails me excitedly with news that the project has now identified four new drugs).

I set out to find real, not speculative, innovation – systems change that I could touch, feel, smell – and find it I did. I haven't covered every system that needs reform, or all the innovators challenging the systems I did choose to cover, nor even found space for all the people I visited on my journey (from Steven Dring and Richard Ballard, who've turned an old tube station in Clapham into a sustainable hydroponic farm, to North Carolina surgeon Kevin Lobdell re-imagining intensive care by learning from NASCAR pitstops). But rest assured, those challengers exist in nearly every walk of life.

And so my experience tells me that we have the tools and ingenuity we need to remake our world. It's also taught me that the problem is the solution – bringing with it the very real hope that the structural inequalities our societies suffer can begin to be bridged. Want to fix agriculture? Engage with the poorest farmers. Want to fix education? You'll find your answer in the worst served students and the most embattled teachers. Broken cities can be reborn through connecting their most neglected citizens. Democracy starts working when those most marginalised are involved in its operation. Those who can't afford to pay for the existing energy system are the ones to own the new one. Indeed, a democratised energy system can help neglected communities invest in the very things their governments may have forgotten to – jobs, schools, hospitals, opportunity. We can, if we choose, thank our old systems for their hard work and pass the baton to some new kids on the block. It might not be easy, as all those I've visited will attest, but it'll be far easier than dealing with the consequences of carrying on the way we are.

The author William Gibson famously said, 'The future is already here – it's just not evenly distributed.' What I've learnt is that many futures are already here – and now we have to decide which ones to distribute. Of course some of the ideas and projects in this book may perish, but, as the great German, Johann Wolfgang von Goethe, once observed, 'Daring ideas are like chessmen moved forward; they may be defeated, but they start a winning game.'

We can build a new operating system for our societies. If you say it is impossible, I invite you to talk to Boston's Jamie Heywood, Delhi's Samir Brahmachari, Jharkhand's Sudhir Paswan, Güssing's Peter Vadasz and Reinhard Koch, Detroit's Ashley Atkinson, the good people of Porto Allegre or Lincoln's Carl Jarvis. Why not talk to them over a beer cooled in an engine that runs on air invented by Peter Dearman?

All of them were told they were crazy, that what they imagined could not be done.

All of them did it anyway.

ACKNOWLEDGEMENTS

My beloved Caroline told me that when I wrote my first book I was, towards the end, largely awful to be around. My biggest thanks must therefore go to her for not only supporting me to write a second, but reading every word several times and giving invaluable feedback. She hates attention so I'll stop there, but I could go on for many paragraphs about her virtues, paragraphs she would put a line through.

Thanks to my editor Mark Ellingham at Profile Books, whose analysis is always spot on, succinct, generous in intent and refreshingly exact. It's a rare combination of good nature and good judgement – qualities shared by many of his colleagues at Profile. Mark has also been enormously patient. This book came in late, largely down to the arrival of a baby, Emmett, in our midst, and Mark was incredibly understanding and supportive.

On the subject of babies I need to say thanks to all of Emmett's grandparents. For everything. When you have kids you suddenly realise just how much your parents loved (and still love) you, and it's something quite staggering to realise in later life. On a practical note, Emmett's maternal grandparents Peter and Margaret Smith also reviewed the section on education (Peter's a brilliant teacher) and have done some sterling grandparenting, which has been invaluable in helping our family find its new normal (and this book get finished within the decade).

Other reviewers included the incisive Christopher Hughey, Claire Marshall (a demon writer on the sharing economy; watch out for her), the polymathic soon-to-be-household-name Dr Kat Arney (whose comments on the healthcare chapters were incredibly useful), the irrepressible and smiley Ben Marchant, my charming niece Charlotte Stevenson and my neighbour and dear friend Paul Milnes. (He did distract me a bit too much with games of Uxbridge English Dictionary but did also selflessly sacrifice his sobriety to help me celebrate finishing the manuscript.) Also in the list of reviewers is Ed Gillespie who could have easily been in the book: a marine biologist turned businessman who's been at the forefront of imagining better for twenty years. Ed says this great thing. 'If you want to subvert the status quo you need to have more fun than they are having – and let them know while you're doing it …' He lives up to that, and inspires me daily.

My agent Charlie Viney deserves a special mention. If I hadn't bumped into him at a New Year's Eve party several years ago, and he hadn't generously listened to me blather on about some ideas I had for books (when he had every right to walk in the direction of the bar), I wouldn't be in the fortunate position I am today of being a published author. I also need to mention Philip Groom, who coincidentally lives on the same street as Charlie. He must also share some of the blame for both this book (and the last) and he knows well why.

I picked many brains but want to especially mention those of the awesome Professor Peter Byck at Arizona State University, Amanda Ravenhill of Project Drawdown, Caroline Julian at ResPublica, health futurologists Dr Pritpal Tamber and Maneesh Juneja, sustainability deity John Elkington, and Professor Anthony Chalmers at the Institute of Cancer Sciences, University of Glasgow – who all were able to shine light into the darkness with an ease that comes from really owning their subject. Three other people who excel in their professions are songwriter Andy Partridge, who generously let me quote from one of his songs; Nikky Twyman, the most diligent of proofreaders; and Harry Hatson, who designed the fabulous cover of this book. Thank you all.

Several people helped me with access to my interviewees and/or the logistics of my visits, sharing their contacts or time, so thanks to

Margot Carlson Delogne at PatientsLikeMe, the tireless Caroline Tecks at the Dearman Engine Company, Roswitha Gruber at the Güssing's European Centre for Renewable Energy, participatory budgeting guru Professor Yves Cabannes, Claudinéai Jacinto at City Hall in Belo Horizonte and Sally Newman at Eos Education. Without them half of what you've just read would never have been written.

On the matter of things never written, special thanks must go to those people who took time to meet me on my journey but never ended up in the book. Iin particular, Gordon Glass, who accompanied me to Brussels to introduce me to the campaign for a United Nations Parliamentary Assembly; Steven Dring and Richard Ballard, the co-founders of Growing Underground (Clapham's sustainable hydroponic farm in a disused tube station); the good people of Repowering London who showed me some of their solar panels on Brixton's rooftops; and surgeon Kevin Lobdell, Director of Quality at the Sanger Heart and Vascular Institute in Charlotte, North Carolina – a champion of Patient-Centred Transformational redesign who's attempting hospital-centred systems change. Kevin, you were an inspiration. Also thanks to Richard Gendall Brown, who unravelled the world of Blockchains for me – a subject I hope to cover in a further volume.

Mention must also be made of Tom Osbourne, who helped me map out hundreds of innovations across the globe, from which the selection covered here was culled. Since we did that, the wonderful people at Atlas of the Future have created a much better version online – and I must thank them for constant inspiration and friendship (and for very kindly asking me to join the team). Cathy Runciman, in particular, is a role model for getting stuff done, and thank you Lisa Goldapple for making whatever it is you're doing potentially illegal but a lot of fun.

My colleagues had to be patient, often finding me unavailable because I was travelling or writing, so fulsome thanks to everyone associated with my network of thinkers, which shares a name with this book: *We Do Things Differently*. Notable among them is Jack Milner, who is unwittingly responsible for my whole career. David Addison who runs the Virgin Earth Challenge, has also had to put up with me being AWOL for a few too many meetings as well. Thanks for understanding, David. Thanks also to all my friends at the London Speaker Bureau, who are a joy to work with, with shout-outs to Tom Kenyon-Slaney, the irrepressible Maria Franzoni and Sian Jones.

A few years ago I came up with the idea for the League of Pragmatic Optimists (LOPO). I got distracted and must therefore thank Chloe Dyson, Tom Mansfield, Kate Wyatt and Emily Brett for reinvigorating the whole shebang. And a more general thanks to everyone involved in LOPO everywhere. Your belief in a better future is the fuel we need.

Penultimately, and in no particular order: Amy Vaughan-Spencer for the buttock-kneading, Andy Ross for the devils, gods and fairy tales, Ian Faragher for delaying The Pig, Richard Bremner for reminding me of the importance of breakfast, Vicky Long for being Vicky Long, Tony Lovell for the love with two Ls, Pippa Heath for happiness embodied, and Catherine and Malcolm Patterson for illuminating the soul.

The final word must go to all those I interviewed and visited on my travels, who dare to look at the status quo and say, 'I can fix that.' I salute you. You changed me, you inspired me, and I hope I've done you justice.

INSPIRATIONS

If you've been inspired by the people in this
book, you might be interested in:

THE ATLAS OF THE FUTURE

www.atlasofthefuture.org

The Atlas of the Future aims to democratise the future: to raise the profiles of the people and projects working to create a better world. Everyone should be able to understand and engage in the topics that affect us all. We only choose projects that are real, innovative, with long-term vision and committed to lasting positive impact.

To get in, the Atlas projects have to be:

1. Real. That means they aren't dealing with the probabilities of futurology, the stuff of science fiction or in the research stages. They are really happening.
2. Innovative. They bring a creative element or unique contribution to solving the challenges facing humanity.
3. Have a long-term vision. The Atlas is not about one-off, flash-in-the-pan-ideas, but a real dedication to the future.
4. Committed to lasting positive impact. These are the innovations that find original solutions to the world's problems, however big or small, and then keep on contributing to the greater good.

THE LEAGUE OF PRAGMATIC OPTIMISTS

www.leagueofpragmaticoptimists.org

Designed as a meeting place where people who want to make the world better can meet, generate ideas and projects, get inspiration and a recharge, the League of Pragmatic Optimists is a place you can find collaborators and have your neurons tickled in the cause of improving the story of humanity. It's a place for diverse people and ideas to bump into each other and create (and deliver) projects that aim to improve things. Local chapters meet at regular intervals to free themselves from day-to-day organisational or mental silos, and network in an inspiring atmosphere of do-ers.

We have eight simple principles and a lot of fun.

NOTES

1 MY BROTHER'S KEEPER

The early origins of our understanding of ALS as a condition can be found in Lewis P. Rowland's article 'How Amyotrophic Lateral Sclerosis got its name; the clinical-pathologic genius of Jean-Martin Charcot' in the March 2001 edition of *JAMA Neurology*. You can find Jamie Heywood's moving announcement of his brother's death on Stephen's PatientsLikeMe.com public profile (username: ALSking). The write-up of the original Celebrex trial that gave the Heywood family (and the ALS community) false hope ('Cyclooxygenase 2 inhibition protects motor neurons and prolongs survival in a transgenic mouse model of ALS') is in the Sept 2002 edition of *Annals of Neurology*.

Nature's write-up of the 'big scathing scandal' ALS TDI uncovered is in its Aug 2008 issue while ALS TDI's dispiriting account of re-running previous ALS trials is detailed in 'Design, power, and interpretation of studies in the standard murine model of ALS', published in 2008 in the journal *Amyotrophic Lateral Sclerosis* (the first issue of that year).

John Ioannidis' seminal paper 'Why most published research findings are false' can be found in *PLOS Medicine* (Aug 2005). I also quoted from two excellent profiles of his work: 'Lies, damned lies, and medical science' by David Freedman in the Nov 2010 issue of the *Atlantic*, and Julia Belluz's article 'John Ioannidis has dedicated his life to quantifying how science is broken' on vox.com.

The *Economist*'s criticism of the selective publication of drug trials (referencing Pfizer's Reboxetine) can be found at their online

'Clinical trial simulator' (economist.com). Dr Marcia Angell's damning indictment of published clinical research features in the her long (and depressing) review of three books (entitled 'Drug Companies & Doctors: A Story of Corruption') that appeared in the *New York Review of Books* in Jan 2009. I drew the statistic that 'dodgy research wastes over $100 billion a year' from 'Avoidable waste in the production and reporting of research evidence,' by Iain Chalmers and Paul Glasziou, printed in *The Lancet*, June 2009.

Jamie's blistering keynote to the Drug Information Association's annual conference can be found on YouTube (titled: 'Keynote presentation at DIA 2014 50th annual meeting'). The shocking rise of deaths from preventable medical errors he references in that talk is documented in various sources, including the book *To Err is Human: Building a Safer Health System* by Linda T. Kohn, Janet M. Corrigan and Molla S. Donaldson, published by the National Academies Press in 2010, John James' paper 'A new, evidence-based estimate of patient harms associated with hospital care' in the Sept 2013 edition of the *Journal of Patient Safety* and the 'Leading causes of death' page on the US Centers for Disease Control and Prevention's website (www.cdc.gov).

2 PATIENTS LIKE ME

You can hear Dave deBronkart's story of surviving cancer and becoming a patient advocate in his April 2011 TEDx Maastricht talk (available on ted.com) and find out more about his work on epatientdave.com. Tom Ferguson's argument that 'people provide their own illness care between 80% and 98% of the time' is drawn from his July 1985 article 'Medical self-care: the seven rules for better health,' for *Mother Earth News*, while his call for 'a new cultural operating system for healthcare' is outlined in his 2007 white paper (co-authored with the e-Patient Scholars Working Group), 'E-patients: how they can help us heal healthcare' (available at e-patients.net).

Surprising revelations about the sexual preferences and spelling abilities of various subsets of the US online dating population are from the okcupid.com blog (oktrends), notably the posts '10 charts about sex' and 'The best questions for a first date', both written by okcupid founder and data-geek Christian Rudder. Proof that people who keep food diaries are more likely to lose weight is available from many sources, but a particularly large study, 'Weight loss during the intensive intervention phase of the weight-loss maintenance trial', published in the Aug 2008 edition of the *American Journal of Preventative Medicine,* is one of the most convincing.

Insights into the veracity of patient records were drawn from 'Accuracy and completeness of electronic patient records in primary care' by Azeem Majeed, Josip Car and Aziz Sheikh (printed in volume 25 of *Family Practice*), 'Health-care providers want patients to read medical records, spot errors' by Laura Landro in the 9 June 2014 edition of the *Wall Street Journal,* 'Impact of electronic health record systems on information integrity: quality and safety implications' by Sue Bowman in the Fall 2013 edition of *Perspectives in Health Information Management* and Julie Cradock, Alexander Young and Greer Sullivan's paper, 'The accuracy of medical record documentation in schizophrenia' (which references the issue of racial prejudice) published in the *Journal of Behavioral Health Services & Research* in Nov 2001. Mary Kerswell's battle with the authorities over the accuracy of her medical records was reported in many places including on the BBC (see 'Medical notes request prompts woman's "arrest" in handcuffs', BBC News, 21 Dec 2012) while the hopeful role of patients improving their records (when they can get access to them) is detailed in 'How patients can improve the accuracy of their medical records' by Prashila M. Dullabh and colleagues in the Oct 2014 edition of the journal *eGEMs.*

The bijou study that gave ALS sufferers hope there may be a therapeutic benefit to taking lithium carbonate (and prompting many to start their own 'off-label' experiments) is found in the *Proceedings of the National Academy of Sciences of the United States of America* (Feb 2008) and is titled 'Lithium delays progression of amyotrophic lateral sclerosis'. The subsequent abandonment of the National Institute of Neurological Disorders and Stroke (NINDS) large-scale trial is announced in 'Clinical trial testing lithium in ALS terminates early for futility' in *Lancet Neurology,* May 2010. The figure of $12 million for the average cost of a NINDS large-scale double-blind trial comes from *The Lancet:* 'Effect of a US national

Institutes of Health programme of clinical trials on public health and costs' by S Claiborne Johnston, John D Rootenberg, Shereen Katrak, Wade S Smith and Jacob S Elkins published in April 2006. Heralding PatientsLikeMe's ability to compete with such trials, it was *Business 2.0* that judged Jamie and co. to be at the helm of one of '15 companies that will change the world' in Aug 2007. PatientsLikeMe's discovery of a larger incidence of compulsive gambling in those taking medication for Parkinson's was published in the April 2009 edition of *Movement Disorders*, while PLM's research into the link between the severity of multiple sclerosis symptoms and the onset of menopause is published in *Multiple Sclerosis and Related Disorders* (Volume 4, Issue 1, Jan 2015). The surprising and hopeful observations of epilepsy sufferers on PatientsLikeMe – following a survey of them – was written up in the Jan 2012 edition of *Epilepsy and Behaviour* ('Perceived benefits of sharing health data between people with epilepsy on an online platform'), findings that prompted co-founder Ben Heywood to say two months earlier, as he gave a presentation for TEDx Cambridge (Massachusetts), 'If I could create a drug that could do that I'd be a very rich man.' The suggestion that 'Patient engagement is the blockbuster drug of the century' comes from an article with that title by Dave Chase for *Forbes*, published in Sept 2012.

Research showing that patients actively involved in their own care spend less time in hospital, manage their conditions better, are subject to fewer medical errors and are more likely to engage productively with their healthcare providers (who they will have a higher opinion of) was drawn from several sources, including the paper, 'Is patient activation associated with future health outcomes and healthcare utilisation among patients with diabetes?' by Carol Remmers and colleagues in the *Journal of Ambulatory Care Management* (2009) and the AARP Public Policy Institute's report 'Chronic Care: A Call to Action Health for Reform', which references the work of Hibbard, Mahoney, Stock, and Tusler. The figure of a 17% reduction in the financial burden suffered by healthcare systems that engaged patients can confer comes from a large study done in Minnesota and written up by Judith H. Hibbard, Jessica Greene and Valerie Overton in the Feb 2013 edition of *Health Affairs* ('Patients with lower activation associated with higher costs').

3 BUG IN THE SYSTEM

You can find data on just how bad the tuberculosis situation is by visiting the website of the World Health Organisation. The growing scale of the problem, in light of the pharmaceutical industry's poor performance in developing new drugs for the disease is outlined in 'Addressing the threat of drug-resistant tuberculosis: a realistic assessment of the challenge' by Robert Giffin and Sally Robinson (National Academies Press, 2009).

I drew my figures on the projected growth of the market for anti-obesity drugs from Infiniti Research's Nov 2014 report 'Global Anti-Obesity Drugs Market 2015–2019', while the eye-watering $2.6 billion figure for the cost of taking a new drug to market comes from Tufts University's 'Innovation in the pharmaceutical industry: new estimates of R&D costs' – research conducted by the university's Centre for the Study of Drug Development and published in the May 2016 edition of the *Journal of Health Economics*. Eroom's Law, describing the ongoing decline in efficiency in the drug development is discussed in 'Diagnosing the Decline in Pharmaceutical R&D Efficiency' by Jack W. Scannell, Alex Blanckley, Helen Boldon and Brian Warrington, which can be found in the March 2012 of *Nature Reviews Drug Discovery* while Professor Chas Bountra's observation that the current system of drug development is 'wasting resources and careers' was aired as part of the BBC Radio 4 programme 'The end of drug discovery?' broadcast on 22 May 2012 and still available, at the time of writing, on the BBC website. The monster revenues that came to Pfizer from sales of its anti-cholesterol pill Lipitor are hard to pin down exactly. I took the $141 billion number from the article 'The Best-Selling Products of All Time', by Vince Calio, Thomas C. Frohlich and Alexander E.M. Hess published in *Time* (and elsewhere) in May 2014, the year after Lipitor's patent expired. Trawling around other sources it's a defensible number and almost certainly in the right ballpark. The reported top-spot for the Sony Playstation comes from the same article.

Rohit Malpani's withering attack on that $2.6 billion figure from Tufts is found in 'Médecins Sans Frontières Response to Tufts CSDD Study on Cost to Develop a New Drug', released 18 Nov 2014, while the more hopeful costs for finding new drugs he quotes, based on the drug development work of the Drugs for Neglected Diseases Initiative, come from 'Research & Development for Diseases of the Poor: A 10-Year Analysis of Impact of the DNDi Model' published in Dec 2013. Accusations of big pharma companies dumping antibiotic compounds into the environment and hence exacerbating the problem of antibiotic resistance are made in 'Bad Medicine: How the pharmaceutical industry is contributing to the global rise of antibiotic-resistant superbugs' – a June 2015 document prepared by SumOfUs based on research by Changing Markets and Profundo.

Dodgy practice by large pharmaceutical companies is well reported – there is a wealth of depressing research out there. The quotes and figures I used in the manuscript were drawn from the Jan 2014 paper 'Financial Conflicts of Interest in Medicine,' by the University of California's Rady School of Management (this specifically dealing with the effectiveness of bribery on doctors) and two BBC News reports: 'GlaxoSmithKline fined $490m by China for bribery' (19 Sept 2014) and 'Pharmaceutical industry gets high on fat profits' (6 Nov 2014).

Figures on how much antimicrobial resistance will cost the USA come from the 'Antimicrobial resistance' page of the website for the Infectious Diseases Society of America. The UK Government's review of antimicrobial resistance has its own website (amr-review. org), which outlines the reasoning behind the worry that, by 2050, drug-resistant infections will kill more people each year than cancer, leading the review's chair, Jim O'Neill, to tell the BBC News (14 May 2014) that the pharma industry is heading for a 'watershed' moment. PriceWaterhouseCoopers' 2007 report 'Pharma 2020: The Vision' raises the worry that 'no country will be able to meet the healthcare needs of its inhabitants by 2020' if we carry on the way we are, while Dr Margaret Chan's stark warning that 'A post-antibiotic era means, in effect, an end to modern medicine as we know it' is quoted in a May 2014 blog post by Dr Tom Frieden for the US Centres for Disease Control, titled 'The end of antibiotics. Can we come back from the brink?'

Lincoln Stein's analogy that genome annotation is not unlike interpreting ancient religious texts is found in an article he wrote for the July 2001 issue of *Nature Reviews Genetics*, 'Genome annotation: from sequence to biology'.

A detailed write-up of the Connect2Decode crowdsourced genome annotation project that Rohit took part in so successfully was published in *PLoS One* in July 2011: 'Crowd sourcing a new paradigm for interactome driven drug target identification in Mycobacterium tuberculosis'. The academic disdain with which it was met can be found (in part) in the article 'India's tuberculosis genome project under fire,' by K. S. Jayaraman in *Nature* (9 June 2010).

The University of British Columbia's investigation into herring flatulence was brought to my attention by an article written by Celeste Biever for *New Scientist*, published in Nov 2003 ('Fish farting may not just be hot air').

You can read about the Mumbai Dabbawalas on their website, mumbaidabbawala.in.

Details of the toxic side effects of the TB drug Rifampin (notably acute liver disease) were drawn from the US National Library of Medicine's 'drug record' for the product.

India's poor showing re: gender equality is laid bare in the *Global Gender Gap Report* from the World Economic Forum; I quoted figures from the 2014 edition. The statistic that only 27% of females participate in the labour market in India comes from the World Bank (see data.worldbank.org for a host of stats on every nation).

GSK's CEO Andrew Witty's admission that the 'blockbuster model' that drives most pharma profits is flawed because it relies on 'finding a needle in a haystack right when you need it' came from a report in *The Economist* ('Triple therapy: Andrew Witty of GlaxoSmithKline has a three-part prescription for the pharmaceuticals giant,' 14 Aug 2008). Witty's observation 'if you stop failing so often you massively reduce the cost of drug development' is reported in an article by Ben Hirschler for *Reuters* ('GlaxoSmithKline boss says new drugs can be cheaper').

Rules for using 'the Malaria Box' are spelled out at mmv.org/malariabox.

Rohit's Twitter revelation that 'I figured out how a cheap drug that is used to treat diabetes can be used for TB' relates to his paper (co-authored with Samir), 'Metformin as a potential combination therapy with existing front-line antibiotics for Tuberculosis', published in the *Journal of Translational Medicine*, while the more expensive, labour-intensive and mouse-killing study that came to much the same conclusions can be found in *Science Translational Medicine*: 'Metformin as adjunct anti-tuberculosis therapy'. Details of the amount of money being spent on OSDD is on its website (see their FAQ, under the question 'Who is funding OSDD?').

Swami Vivekananda's advice to 'Be obedient and eternally faithful to the cause of truth, humanity, and your country' (along with nearly everything else he wrote) can be found in *The Complete Works of Swami Vivekananda* (available from various sources online).

4 RICE WARS

Colourful attacks on SRI can be found in the articles 'Curiosities, nonsense, non-science and SRI' by J.E. Sheehy, Thomas R. Sinclair and K.G. Cassman, and 'Does the system of rice intensification outperform conventional best management?' by A.J. McDonald, P.R. Hobbs and S.J. Riha. Both can be read in *Field Crops Research* (volumes 91 and 96 respectively). There's further disparagement from Sinclair in his article 'Agronomic UFOs waste valuable scientific resources' for *Rice Today* (July 2004) and from Sheehy and colleagues in 'Fantastic yields in the system of rice intensification: Fact or fallacy?' (also in *Field Crops Research*, this time volume 88).

I found statistics outlining Jharkhand's poverty in a presentation by the Indian Planning Commission on its five-year plan (2012–17) for the region, available at planningcommission.nic.in. Figures on the extent of irrigation in the state came from the website of Jharkhand's State Agricultural Management & Extension Training Institute. Information on levels of worldwide food production for various years are drawn from the Food and Agriculture Organisation (FAO) of the UN's statistics website (faostat.fao.org) and Chapter 3 of 'Diet, nutrition and the prevention of chronic diseases', a joint report from the UN FAO/ WHO published in 2003. Data outlining the decline in famine comes from the ever-useful faostat website and 'A History of global famine deaths' in *The Economist*.

It was magician Penn Jillette who described Norman Borlaug as 'the greatest human being you've probably never heard of'. I drew my brief biography of Norman (including the controversy surrounding his work) from three main sources: 'Green Giant', by Mark Stuertz in the *Dallas Observer* (5 Dec 2002), Borlaug's obituary in the *Guardian*, and 'Against the Grain on Norman Borlaug' by Leo Hickman in the same newspaper (15 Sept 2009). Alexander Cockburn's blistering denouncement of Borlaug and the Green Revolution can be found on the website counterpunch.org, in an article entitled 'Al Gore's Peace Prize' published 13 Oct 2007.

Depressing news about the state of India's and world's freshwater resources come from a host of places, including the Jet Propulsion Laboratory (see: 'Third of big groundwater basins in distress', June 16 2015) and Columbia University's Water Centre at the Earth Institute, on whose website you will also find Lester Brown's book, *Plan B 2.0: Rescuing a Planet Under Stress and a Civilization in Trouble*, available for free. Chapter 3 of that book, 'Emerging water shortages: falling water tables', offers a good summary of our problems and explains why, as Lester warns in his July 2013 article for the *Guardian* ('The real threat to our future is peak water'), the world is facing 'a water-based food bubble'.

I used various newspaper articles reporting on drought and the problems with ever more expensive and deeper wells to fill out the dire picture. These included: 'Dry wells plague California as drought has water tables plunging' by Alison Vekshin for *Bloomberg Business*, 'Beijing closes 6,900 wells to protect underground water' in *China Weekly* (Nov 2014), and 'Asian farmers sucking the continent dry' in *New Scientist*. Stories of rural hardship and despair in India come from National Public Radio's 'Green Revolution: trapping India's farmers in debt' by Daniel Zwerdling (April 14 2009), and 'In India, agriculture's Green Revolution dries up' by Kenneth R. Weiss for the 22 July 2013 edition of the *Los Angeles Times*.

I found stats for farmer suicides in India on the website of the Indian National Crime Records Bureau, specifically the 2014 'Accidental deaths and suicides' report, which dedicates a whole chapter to 'Farmer suicides' – underlining the seriousness of the problem.

The rise in fertilizer use is documented on the International Fertilizer Industry Association's data website (ifadata.fertilizer.org). Fusuo Zhang's research showing how its overuse has brought yields down by as much as 50% in some areas of China, and his warning that if things carry on as they are this could 'cripple Chinese agricultural production', are covered in Natasha Gilbert's article for *Nature*, 'Acid soil threatens Chinese farms' (Feb 11 2010). Data on the extent of, and the science behind, marine 'dead zones' come from the Virginia Institute of Marine Science, whose website has a helpful section on the phenomenon.

You can find a list of all those herbicide-resistant weeds at the International Survey of Herbicide Resistant Weeds (weedscience. org), while you can pick up the worrying and long established picture on soil erosion by reading 'Soil-Erosion and Runoff Prevention by Plant Covers: A Review' by Victor Hugo, Duran Zuazo, Carmen Pleguezuelo in *Agronomy Sustainable Development* (March 2008) and 'Environmental and economic costs of soil erosion and conservation benefits' by David Pimentel and colleagues in *Science* (vol. 267, issue 5201, Feb 1995). A good summary of the problem of soil carbon finding its way into the atmosphere (and some ideas about what to do about it) comes from Rattan Lal of the Carbon Management and Sequestration Centre at Ohio State University, in particular his paper 'Crop residues and soil carbon'. In presentations of his work, Lal says: 'irrespective of the climate debate, soil quality and its organic matter content must be restored, enhanced and improved'.

Father Henri de Laulanié's own account of his experiments can be found in the paper 'Technical presentation of the system of rice intensification, based on Katayama's tillering model' available on the website of the SRI International Network and Resources Centre (where Erika is the boss). I drew my figure on average rice yields (between 4 and 5 tonnes) from the FAO's Rice Market Monitor. Sumant Kumar's whopping yield of 22.4 tonnes per hectare was widely reported (and disputed), including in the *Observer* (see John Vidal's 'India's Rice Revolution' from 16 Feb 2013). You can find a report of Nitish and Rakesh Kumar's world-beating harvests of 'SRI' potatoes in the article

'Agriculture revolution takes shape silently' by Faizan Ahmad for *The Times of India* (22 March 2013). A detailed description of Sumant's SRI-inspired approach (acknowledging his use of fungicide-treated hybrid seeds and small amounts of fertilizer and herbicide) can be found in the June 2012 edition of *Agriculture Today*, written by M.C. Diwakar, Arvind Kumar, Anil Verma and Norman Uphoff; this article is referenced in Achim Dobermann's balanced take on SRI, which can be found on his blog for the International Rice Research Institute in the post 'Another new rice yield record? Let's move beyond it'.

Stats on the number of smallholders worldwide (and the size of their holdings) were taken from the UN FAO, specifically ESA Working Paper no. 14: 'What do we really know about the number and distribution of farms and family farms in the world?' by Sarah K. Lowder, Jakob Skoet and Saumya Singh, and the factsheet 'Family Farmers: Feeding the world, caring for the earth'. Olivier de Schutter's hopeful report to the UN on agro-ecological methods and their benefits can be downloaded from UN Human Rights website.

5 RUNNING ON AIR

The history of Hans Knudsen's Liquid Air, Power and Automobile Company was pulled together from a report in the 26 Aug 1899 edition of the *Cambridge Chronicle*, Archibald Williams' chapter on the company for his 1903 book *The Romance of Modern Invention* (a cracking read, even today), and Royal Feltner's exhaustive website earlyamericanautomobiles.com.

You can find details of Peter Dearman's legal battles on the website of the UK Government Intellectual Property Office ('An application under Section 13(3) by Kevin Bowden in respect of Patent No. 2270629).

Information on Washington University's nitrogen car project that appeared on *Tomorrow's World* can be found on its website (see 'Why Liquid Nitrogen Cars Are Better Than Electric Autos'). The claim that the car ran for the same price per mile as gasoline stacks up, taking information on historical US gasoline prices available from the US Energy Information Administration and the average fuel efficiency of a car sold in America at the time: about 29 miles per gallon (data from the US Department of Transportation).

Hunger and food loss statistics were drawn from the websites of UN FAO and the World Food Programme (notably the UN FAO reports *Food losses and waste in the context of sustainable food systems* and *Food wastage footprint*) and Tim Fox's *Global food: waste not want not* report written for the Institution of Mechanical Engineers (IMechE). The prediction that the world population 'will reach nearly 11 billion people by the end of the century and then level off' comes from research available at the UN Department of Economic and Social affairs, Population Division.

Figures concerning the world market for refrigeration were drawn from two reports: *Cold chain market by type: global trends & forecast to 2019* by Markets and Markets (that's not a typo) and *World commercial refrigeration equipment market* by Freedonia Research, both published in Nov 2014.

I found that '16% of all electricity consumed in the UK goes on cooling' by reading *Sustainably meeting the global food crisis: why we need 'green' cold chains*, a report that accompanied a House of Lords panel discussion on liquid air technology that took place on 14 July 2015. The worrying revelation that India only has 7,000 refrigerated trucks for transporting perishable produce (in a population of 1.2 billion) comes from a presentation to the IMechE given by Pawanexh Kohli, CEO of India's National Centre for Cold Chain Development. Lisa Kitinoja's report, *Exploring the potential for cold chain development in emerging and rapidly industrialising economies through liquid air refrigeration technologies*, is available at liquidair.org. uk. Worldwide numbers of refrigerated trucks were taken from the '2010 assessment of refrigeration, air conditioning and heat pumps' by the Technical Options Committee at the Ozone Secretariat of the UN Environment Programme.

The potential for liquid air technology to deliver £1 billion in annual revenues and 22,000 jobs to the UK is outlined in a report titled *Liquid air in the energy and transport systems: opportunities for industry and innovation in the UK* from the Centre for Low Carbon Futures (a collaboration between the universities of Hull, York, Sheffield, Leeds and Birmingham).

6 INSTANT POWER

Details of Britain's energy storage mix come from 'Energy storage in the UK electrical network: Estimation of the scale and review of technology options' by Ian Allan Grant Wilson, Peter G. McGregor and Peter Hall, published in *Energy Policy* (no. 38, April 2010).

Figures pertaining to electricity access levels in various nations came from a few sources, including the World Bank, the UN Development Programme and the International Energy Agency.

Duke University's study showing the exponential drop in battery prices is 'An evaluation of current and future costs for lithium-ion batteries for use in electrified vehicle powertrains' by David Anderson. This acted as the main source for the wonderful Ramez Naam's blog post 'Energy storage gets exponentially cheaper too', published on 25 Sept 2013. UBS's research suggesting a fourfold drop in battery prices within a decade is to be found in a report titled 'Will solar, batteries and electric cars re-shape the electricity system?' from Aug 2014. The toxicity of lithium-ion batteries is outlined in an April 2013 report 'Application of lifecycle assessment to nanoscale technology: lithium-ion batteries for electric vehicles' by the National Risk Management Research Laboratory at the US Environmental Protection Agency's Office of Research and Development.

7 EDISON'S REVENGE

Arnold Schwarzenegger's remark that 'The whole world should become Güssing' was widely reported at the time, including on the districtenergy.org blog ('Arnold Schwarzenegger bullish on district energy, renewable energy', 24 June 2012).

I drew the figure of €6 million spent on energy in Güssing (prior to its energy transformation) from a presentation called 'A model for regional economic renewal through low carbon innovation' given by Christian Doczekal of Güssing Energy Technologies. Figures for the energy imports of various nations were provided by the World Bank, the House of Commons Library and the US Energy Information Administration.

The stories of Edison, Tesla and Insull are exhaustively documented. I drew together my account and supporting statistics from sources provided by the Institute of Electrical and Electronics Engineers, the Institute for Energy Research, the National Museum of American History and the US National Council on Electricity Policy. The demise of unfortunate maintenance engineer John Feeks was reported in the 12 Oct 1889 edition of the *New York Times* ('Horrifying spectacle on a Telegraph Pole'). I drew on Walter Pagel's 1944 book, *The Religious and Philosophical Aspects of Van Helmont's Science and Medicine* when researching Jan Baptista van Helmont's work with 'gas sylvestre'. The statistic that coal-fed power plants are, on average, only 33% efficient came from the World Coal Association. The contrasting figure of 81.3% efficiency for Güssing's Biomassekraftwerk's came from a detailed write-up of the plant for the book, *Biomass Power for the World: Transformations to Effective Use*, published in May 2015 by Pan Stanford (see the chapter 'Energy from biomass via gassification in Güssing').

8 ENERGY TRILEMMA

The conflicting figures of $500 billion (IEA) and $5 trillion (IMF) for worldwide fossil fuel energy subsidies come from the International Energy Agency's *Redrawing the energy-climate map, world energy outlook special report* and an IMF working paper produced by its Fiscal Affairs Department, 'How large are global energy subsidies?' The more detailed figures on the split of subsidies between fossil fuel and renewables in the UK and US came from the US Energy Information Administration, the 'Digest of United Kingdom Energy' produced by the (now defunct) Department of Energy and Climate Change, and evidence presented to the Commons Select Committee's 2013 enquiry into energy subsidies in the UK. Maria van der Hoeven's view that 'fossil fuel subsidies are an extremely inefficient means of achieving their stated objective' came from a presentation she gave at the Oxford Energy Colloquium on 27 Jan 2015.

Statistics on the cost of air pollution came from several sources, notably 'Economic value of US fossil fuel electricity health impacts' by Ben Machol and Sarah Rizk from the Clean Energy and Climate

Change Office of the US Environmental Protection Agency Region and published in *Environment International* (vol. 52, Feb 2013); 'The Unpaid Health Bill – How coal power plants make us sick,' a March 2013 report by the Health and Environment Alliance; 'The Mortality effects of long-term exposure to Particulate Air Pollution in the United Kingdom', a 2010 report by the Committee on the Medical Effects of Air Pollutants for the UK government; and an ongoing collaboration between Beijing University and Greenpeace monitoring Chinese air pollution. Figures on the projected costs of climate change came from *American Climate Prospectus: Economic risks to the United States*, published Oct 2014 by Risky Business, the research house set up by Bloomberg, Paulson and Steyer.

9 MAKE WAY FOR THE ENERNET

The Edison Electric Institute report I quote from is *Disruptive Challenges: Financial Implications and Strategic Responses to a Changing Retail Electric Business*, while its anti-solar PR activities were revealed by the Energy and Policy Institute, who provide a detailed breakdown of the EEI's PR spending on their website.

My description of Germany's ongoing transition to a low-carbon nation (they even have a word for it: *energiewende*) drew on the work in Caroline Julian's report for ResPublica, *Creating local energy economies, lesson from Germany*, figures from the Fraunhofer Institute for Solar Energy Systems and the article 'Germany to shut down coal-fired plants,' *Reuters*, 1 July 2015.

Coverage of negative wholesale electricity prices on sunny days in Germany and the UK came from the *Independent* ('People in Germany are now being paid to consume electricity' by Doug Bolton, 11 May 2016) and the *Financial Times* ('UK power prices go negative as renewables boom distorts market' by Pilita Clark, 20 May 2016).

The story of Bob Metcalfe literally eating his own words comes from a profile of the man in *Wired* – 'The Legend of Bob Metcalfe' by Scott Kirsner. You can watch Bob outline his Enernet concept at various places online. I drew my quotes from a presentation he gave to the

French Network Operators Group in June 2013 which can be found on its website. You can watch Liu Zhenya's presentation pertaining to a Global Energy Internet by visiting the website of the IEEE, and the archives of their General Meeting in July 2014. Nicola Shaw's and Michelle Hubert's quotes on the UK's coming energy transition were reported by the BBC ('Smart energy revolution could help to avoid UK blackouts' by Roger Harrabin, 31 Aug 2016). The discussion of Saudi Arabia's shift away from its dependence on oil revenues was pulled together for a variety of sources including 'Sheikh Yamani predicts price crash as age of oil ends' by Mary Fagan in *The Daily Telegraph* on June 25 2000, 'The 2 trillion project to get Saudi Arabia's economy off oil' by Peter Waldman for *Bloomberg* on 21 April 2016 and 'Saudi Arabia Plans $2 Trillion Megafund for Post-Oil Era' by John Micklethwait, Glen Carey, Alaa Shahine and Matthew Martin, also for *Bloomberg*, three weeks earlier. News of the rock-bottom solar energy prices now coming to the UEA is from Neil Halligan's article 'The era of solar energy has finally arrived in the GCC' for *Arabian Business*, 10 May 2015.

Stories of the technologies that make fuel from the sky were drawn, in part, from 'Audi's water-based green diesel provides alternative to electro-mobility' by Tereza Pultarova in *Engineering and Technology* magazine, 27 April 2015; the paper 'Nanostructured transition metal dichalcogenide electrocatalysts for CO_2 reduction in ionic liquid' by Mohammad Asadi and colleagues, published in *Science*, vol. 353 on July 29 2016; and a meeting summary of the event 'Solar Fuels and Artificial Photosynthesis: Global initiatives and opportunities' held at the Royal Society of Chemistry on 17 May 2012.

10 ALWAYS BET ON THE TORTOISE

Detroit's upsetting crime statistics were pulled from the UN Office of Drugs and Crime, while my summary of the city's decline drew on 'From Motor City to Motor Metropolis: how the automobile industry reshaped urban America' by Thomas J. Sugruw, *The Ku Klux Klan in the City, 1915–1930* by Kenneth T. Jackson (Oxford University Press, 1967), and the article 'Detroitism' by John Patrick Leary for *Guernica*, 15 Jan 2011.

Statistics concerning obesity came from the Centers for Disease Control and Prevention, research by the Department of Family Medicine and Public Health Sciences at Wayne State University School of Medicine, and the paper 'Direct medical cost of overweight and obesity in the United States: a quantitative systematic review' by Adam Gilden Tsai, David F. Williamson and Henry A. Glick, in *Obesity Reviews*, Jan 2011. I took Barry Popkin's quote about the power of food culture from the article 'Giving the poor easy access to healthy food doesn't mean they'll buy it' by Margot Sanger-Katz in the *New York Times*, 8 May 2015.

Those 'numerous studies' that show 'farmers have a much healthier diet than their supermarket-sourcing neighbours' include 'The influence of social involvement, neighborhood aesthetics, and community garden participation on fruit and vegetable consumption' led by Jill Litt and published in the *American Journal of Public Health*, Aug 2011; 'A dietary, social and economic evaluation of the Philadelphia urban gardening project' by Dorothy Blair, Carole Giesecke and Sandra Sherman, published in the *Journal of Nutrition Education* (July–Aug 1991); and 'Fruit and vegetable intake among urban community gardeners' by Katherine Alaimo and colleagues, also in the *Journal of Nutrition Education*, (March–April 2008).

Michigan State University's Center for Regional Food Systems assessment that it is possible to grow 'roughly three-quarters of vegetables and nearly half of fruits' consumed by Detroiters within the city limits can be found in its report *Growing Food in the City: The Production Potential of Detroit's Vacant Land* by Kathryn Colasanti, Charlotte Litjens and Michael Hamm. The UN report on urban farming in Rosario, Argentina is titled *Urban and Peri-urban Agriculture in Latin America and the Caribbean: Rosario* and is available on the UN FAO website.

11 HOW TO MAKE POLITICIANS POPULAR

The use of torture by the Brazilian military dictatorship is laid bare in *Torture in Brazil* (University of Texas Press, Jan 1998). Arguments that Dilma Rousseff's impeachment is 'a parliamentary coup brought about by corrupt politicians' can be found in Pedro Zambarda's article 'Every vote for Dilma Rousseff's impeachment was a vote for corruption' in the *Huffington Post*, 2 Sept 2016. Many of the stats in the chapter on the state of worldwide democracy (and individual democracies) were courtesy of *The Economist*'s Democracy Index, while the majority of figures concerning corruption came from Transparency International's Corruption Perceptions Index. US millennials fading faith in democracy is outlined in the paper, 'The Democratic Disconnect' by Roberto Stefan Foa and Yascha Mounk, published in the July 2016 edition of the *Journal of Democracy*.

Philosopher Roger Scruton's views on democracy can be found in his article 'Is democracy overrated?' for *BBC News Magazine*, 9 Aug 2013. The discussion of inequality drew on many sources, including 'A Guide to Statistics on Historical Trends in Income Inequality' by Chad Stone, Danilo Trisi, Arloc Sherman and Brandon Debot at the Centre on Budget and Policy Priorities; Allianz's Global Wealth Report 2015, 'Wealth in Great Britain 2012 to 2014' from UK Office for National Statistics; Credit Suisse's 'Global Wealth Report 2014'; and GINI rankings from the World Bank.

Statistics exposing Americans dissatisfaction with their institutions were taken from Gallup's 'Confidence in Institutions' surveys, while the correlation between the level of satisfaction citizens feel about the way their democracy works and their propensity to vote is made in 'Low Electoral Turnout: An Indication of a Legitimacy Deficit?' by Kimmo Grönlund and Maija Setälä – a paper prepared for the European Consortium for Political Research in April 2004.

The story of Porto Alegre's early experiments with participatory budgeting (along with many supporting stats) were pulled from a wide range of sources, including 'The Power of Ambiguity: How Participatory Budgeting Travels the Globe' by Ernesto Ganuza and Gianpaolo Baiocchi in the *Journal of Public Deliberation*; 'Budgets for the People, Brazil's Democratic Innovations' by *Paolo* Spada and Hollie Russon Gilman in *Foreign Affairs* (11 March 2015); 'Hope for democracy – 25 years of participatory budgeting worldwide,' published by the In Loco Association (which also includes Olívio Dutra's words 'democracy's problems are solved with more democracy'); 'From Clientelism to Participation: the story of participatory budgeting in Porto Alegre', an article by Brendan

Martin available at publicworld.org; *Militants and Citizens: the politics of participatory democracy in Porto Alegre* by Gianpaolo Baiocchi (Stanford University Press, 2005); 'Participatory Budgeting in Porto Alegre: toward a redistributive democracy' by Boaventura de Sousa Santos in *Politics & Society* (Dec 1998); 'Contribution of Participatory Budgeting to provision and management of basic services: municipal practices and evidence from the field' by Yves Cabannes, written for the International Institute for Environment and Development; along with various documents prepared by the city to support the process.

Proof of Belo Horizonte's success in encouraging participation from the city's most disadvantaged groups comes from 'Participatory Democracy in Brazil and Local Geographies: Porto Alegre and Belo Horizonte compared' by Terence Wood and Warwick E. Murray in the *European Review of Latin American and Caribbean Studies* (Oct 2007). The figure that '80% of the city's population are living within half a kilometre of infrastructure financed by the participatory budget' is sourced from *Participatory Budgeting Worldwide*, a Nov 2013 report for the German Federal Ministry for Economic Cooperation and Development. Participatory budgeting's positive effects on infant mortality in Brazil are outlined in 'Power to the People: The Effects of Participatory Budgeting on Municipal Expenditures and Infant Mortality in Brazil' by Sónia Gonçalves published in *World Development*, 2014.

The hopeful conclusion that cities which stick with participatory budgeting reap many benefits that without the process do not (as well as a good few stats I use) came from the World Bank's 2008 report, *Brazil: toward a more inclusive and effective participatory budget in Porto Alegre* and *Does participatory budgeting improve decentralised public service delivery?* by Diether Beuermann and Maria Amelina, a 2014 report for the Inter-American Development Bank (this also includes the reference to PB's ability to increase citizens' willingness to pay tax). The political dividend for parties adopting PB is explored in 'The Economic and Political Effects of Participatory Budgeting' by Paolo Spada, prepared for the 2009 Congress of the Latin American Studies Association in Rio de Janeiro, June 2009. You can find evidence of the Obama administration's cautious endorsement of PB in the White House's Second Open Government National Action Plan, published in Dec 2013.

Olívio Dutra summation that 'The participatory budget combines direct democracy with representative democracy, an achievement that should be preserved and valued' comes from 'Quand les habitants gèrent vraiment leur ville' in *Le Budget Participatif: l'expérience de Porto Alegre au Brésil* by Tarso Genro and Ubiratan de Souza, Éditions Charles Léopold Mayer (1998).

If you're interested in learning more about participatory budgeting, I'd recommend two books I have quoted from: Josh Lerner's short but highly readable *Everyone Counts – Could Participatory Budgeting Change Democracy?* (Cornell University Press, 2014) and Hollie Russon Gilman's more academically written *Democracy Reinvented* (Brookings Institution Press, 2016).

12 WORST SCHOOL IN THE COUNTRY

You can find Ken Robinson's download-busting 2006 talk 'Do Schools Kill Creativity?' on ted.com, while the criticisms of his work that I quote come from two blog posts: 'What Sir Ken got wrong,' by Joe Kirby on the Pragmatic Education blog (12 Oct 2013) and 'Ken Robinson rebuttal', by Scott Goodman on the Ed Tech Now blog (Nov 2010).

You can find Ofsted's dim opinion of Hartsholme Academy prior to Carl's appointment in its 2007 School Inspection Report (available from reports.ofsted.gov.uk), while the school's sterling statistical performance compared to the national average is verified at www.compare-school-performance.service.gov.uk.

High Tech High's achievements are trumpeted on their website while the founder Larry Rosenstock's statement that 'I want to be remembered for the quality of my schools, not the quantity' comes from an article partly about Bill Gates' funding of the school on Bloomberg ('Bill Gates gets schooled', 26 June 2006).

The education dividend of David Price's Musical Futures Project is celebrated in a number of places including the initiative's own website but also in *Survey of Musical Futures*, a report from Institute of Education at the University of London by Professor Susan Hallam, and *Musical Futures: a case study investigation*, also for the Institute of Education where Susan Hallam is joined by co-authors Dr Andrea Creech and Dr Hilary McQueen.

Problems with student suicide have been reported widely, including in the *Times Educational Supplement* ('When school stress becomes a matter of life and death' by Richard Vaughan, 23 May 2014); *La Voix des Jeunes* ('Student Suicides in South Korea'); *The Daily Telegraph* ('Chinese school installs 'anti-suicide' barriers before dreaded exam' by Tom Phillips, 21 April 2015); and *University World News* ('Suicides raise concerns at universities' by Mimi Leung, 23 Oct 2013).

Figures for, and some observations on, student disengagement were drawn from: 'The Status Quo: Engaging Schools: Fostering High School Students' Motivation to Learn', by the Committee on Increasing High School Students' Engagement and Motivation to Learn (The National Academies Press, 2004); 'What did you do in School Today?: The Relationship Between Student Engagement and Academic Outcomes' by Jodene Dunleavy, J. Douglas Willms, Penny Milton and Sharon Friesen for the Canadian Education Association (Sept 2012); and more recent student polls by Gallup. The 7% school drop-out rate in the UK is documented in 'Participation in Education, Training and Employment by 16-18 year olds in England' from the UK Department of Education. In the US, the figure (also around 7%) came from a statistical release 'High school dropout rates' prepared by Child Trends (a non-profit research body).

You can find John Hattie's 'Hattie rankings' on his website visible-learning.org. I also drew on his book, *Visible Learning: A synthesis of over 800 meta-analyses relating to achievement*, several of his talks (including 'Why are so Many of our Teachers and Schools so Successful?' given at TEDx Norrköping on 16 Nov 2013), and a profile of him and his methods on the BBC Radio 4 programme *The Educators*. Various other quotes and insights from him (or from others questioning his approaches) were drawn from 'Effective debate: in defence of John Hattie' by Stuart Lord on thelearningintention.net (22 March 2015); 'Hattie makes his move in on early childhood: Key set to act in tandem' by Kelvin Smythe on the Networkonnet blog (April 27 2014); and 'He's not the messiah,' by Darren Evans in the *Times Educational Supplement*, (Sept 14 2012).

I drew the observation that 'collective teacher efficacy' is 'more important in explaining school achievement than socioeconomic status' from 'Toward an organizational model of achievement in high schools: The significance of collective efficacy' by Hoy, Sweetland, and Smith, published in *Educational Administration Quarterly*, issue 38, 2002.

If you haven't found enough information about the reference you are looking for, do contact me via *www.markstevenson.org*